D AF757473

Tidy Finance with R

This textbook lifts the curtain on reproducible finance and shows how to apply theoretical concepts from finance and econometrics by providing a fully transparent R code base. Focusing on coding and data analysis with R, we illustrate how students, researchers, data scientists, and professionals can conduct research in empirical finance from scratch. We start with a beginner-friendly introduction to the tidyverse family of R packages around which our approach revolves. We then show how to access and prepare common open-source (e.g., French data library, macroeconomic data) and proprietary financial data sources (e.g., CRSP, Compustat, Mergent FISD, and TRACE). We present data management principles using an SQLite database, which constitutes the basis for the applications presented in the subsequent chapters. The empirical applications range from key concepts of empirical asset pricing (e.g., beta estimation, portfolio sorts, performance analysis, and Fama-French factors) to modeling and machine learning applications (e.g., fixed effects estimation, clustering standard errors, difference-in-difference estimators, ridge regressions, Lasso, Elastic nets, random forests, and neural networks) and portfolio optimization techniques.

Highlights

1. Self-contained chapters on the most important applications and methodologies in finance, which can easily be used for the reader's research or as a reference for courses on empirical finance.
2. Each chapter is reproducible in the sense that the reader can replicate every single figure, table, or number by simply copy-pasting the code we provide.
3. A full-fledged introduction to machine learning with tidymodels based on tidy principles to show how factor selection and option pricing can benefit from Machine Learning methods.
4. A chapter that shows how to retrieve and prepare the most important datasets in the field of financial economics: CRSP and Compustat. The chapter also contains detailed explanations of the most relevant data characteristics.
5. Each chapter provides exercises that are based on established lectures and exercise classes and which are designed to help students to dig deeper. The exercises can be used for self-studying or as a source of inspiration for teaching exercises.

Chapman & Hall/CRC
The R Series

Series Editors
John M. Chambers, Department of Statistics, Stanford University, California, USA
Torsten Hothorn, Division of Biostatistics, University of Zurich, Switzerland
Duncan Temple Lang, Department of Statistics, University of California, Davis, USA
Hadley Wickham, Posit, Boston, Massachusetts, USA

Recently Published Titles

Javascript for R
John Coene

Advanced R Solutions
Malte Grosser, Henning Bumann, and Hadley Wickham

Event History Analysis with R, Second Edition
Göran Broström

Behavior Analysis with Machine Learning Using R
Enrique Garcia Ceja

Rasch Measurement Theory Analysis in R: Illustrations and Practical Guidance for Researchers and Practitioners
Stefanie Wind and Cheng Hua

Spatial Sampling with R
Dick R. Brus

Crime by the Numbers: A Criminologist's Guide to R
Jacob Kaplan

Analyzing US Census Data: Methods, Maps, and Models in R
Kyle Walker

ANOVA and Mixed Models: A Short Introduction Using R
Lukas Meier

Tidy Finance with R
Christoph Scheuch, Stefan Voigt and Patrick Weiss

Deep Learning and Scientific Computing with R torch
Sigrid Keydana

Model-Based Clustering, Classification, and Density Estimation Using mclust in R
Lucca Scrucca, Chris Fraley, T. Brendan Murphy, and Adrian E. Raftery

Spatial Data Science: With Applications in R
Edzer Pebesma and Roger Bivand

For more information about this series, please visit: https://www.crcpress.com/Chapman--HallCRC-The-R-Series/book-series/CRCTHERSER

Tidy Finance with R

Christoph Scheuch
Stefan Voigt
Patrick Weiss

CRC Press
Taylor & Francis Group
Boca Raton London New York

CRC Press is an imprint of the
Taylor & Francis Group, an **informa** business
A CHAPMAN & HALL BOOK

Designed cover image: Christoph Scheuch, Stefan Voigt and Patrick Weiss

First edition published 2023
by CRC Press
6000 Broken Sound Parkway NW, Suite 300, Boca Raton, FL 33487-2742

and by CRC Press
4 Park Square, Milton Park, Abingdon, Oxon, OX14 4RN

CRC Press is an imprint of Taylor & Francis Group, LLC

© 2023 Christoph Scheuch, Stefan Voigt and Patrick Weiss

Reasonable efforts have been made to publish reliable data and information, but the author and publisher cannot assume responsibility for the validity of all materials or the consequences of their use. The authors and publishers have attempted to trace the copyright holders of all material reproduced in this publication and apologize to copyright holders if permission to publish in this form has not been obtained. If any copyright material has not been acknowledged, please write and let us know so we may rectify in any future reprint.

Except as permitted under U.S. Copyright Law, no part of this book may be reprinted, reproduced, transmitted, or utilized in any form by any electronic, mechanical, or other means, now known or hereafter invented, including photocopying, microfilming, and recording, or in any information storage or retrieval system, without written permission from the publishers.

For permission to photocopy or use material electronically from this work, access www.copyright.com or contact the Copyright Clearance Center, Inc. (CCC), 222 Rosewood Drive, Danvers, MA 01923, 978-750-8400. For works that are not available on CCC, please contact mpkbookspermissions@tandf.co.uk

Trademark notice: Product or corporate names may be trademarks or registered trademarks and are used only for identification and explanation without intent to infringe.

ISBN: 978-1-032-38933-2 (hbk)
ISBN: 978-1-032-38934-9 (pbk)
ISBN: 978-1-003-34753-8 (ebk)

DOI: 10.1201/b23237

Typeset in CMR10
by KnowledgeWorks Global Ltd.

Publisher's note: This book has been prepared from camera-ready copy provided by the authors.

Access the Support Material: github.com/voigtstefan/tidy_finance

Contents

Preface ix

Author biographies xvii

I Getting Started 1

1 Introduction to Tidy Finance 3
- 1.1 Working with Stock Market Data . . . 3
- 1.2 Scaling Up the Analysis . . . 8
- 1.3 Other Forms of Data Aggregation . . . 12
- 1.4 Portfolio Choice Problems . . . 14
- 1.5 The Efficient Frontier . . . 17
- 1.6 Exercises . . . 19

II Financial Data 21

2 Accessing & Managing Financial Data 23
- 2.1 Fama-French Data . . . 24
- 2.2 q-Factors . . . 25
- 2.3 Macroeconomic Predictors . . . 26
- 2.4 Other Macroeconomic Data . . . 28
- 2.5 Setting Up a Database . . . 29
- 2.6 Managing SQLite Databases . . . 32
- 2.7 Exercises . . . 33

3 WRDS, CRSP, and Compustat 35
- 3.1 Accessing WRDS . . . 35
- 3.2 Downloading and Preparing CRSP . . . 36
- 3.3 First Glimpse of the CRSP Sample . . . 41
- 3.4 Daily CRSP Data . . . 45
- 3.5 Preparing Compustat Data . . . 47
- 3.6 Merging CRSP with Compustat . . . 49
- 3.7 Some Tricks for PostgreSQL Databases . . . 52
- 3.8 Exercises . . . 52

4 TRACE and FISD 55
4.1 Bond Data from WRDS 56
4.2 Mergent FISD 56
4.3 TRACE 59
4.4 Insights into Corporate Bonds 60
4.5 Exercises 64

5 Other Data Providers 65
5.1 Exercises 68

III Asset Pricing 69

6 Beta Estimation 71
6.1 Estimating Beta using Monthly Returns 71
6.2 Rolling-Window Estimation 73
6.3 Parallelized Rolling-Window Estimation 75
6.4 Estimating Beta using Daily Returns 78
6.5 Comparing Beta Estimates 80
6.6 Exercises 85

7 Univariate Portfolio Sorts 87
7.1 Data Preparation 88
7.2 Sorting by Market Beta 88
7.3 Performance Evaluation 89
7.4 Functional Programming for Portfolio Sorts 90
7.5 More Performance Evaluation 91
7.6 The Security Market Line and Beta Portfolios 92
7.7 Exercises 97

8 Size Sorts and p-Hacking 99
8.1 Data Preparation 100
8.2 Size Distribution 100
8.3 Univariate Size Portfolios with Flexible Breakpoints 104
8.4 Weighting Schemes for Portfolios 105
8.5 P-hacking and Non-standard Errors 107
8.6 The Size-Premium Variation 109
8.7 Exercises 109

9 Value and Bivariate Sorts 111
9.1 Data Preparation 111
9.2 Book-to-Market Ratio 112
9.3 Independent Sorts 114
9.4 Dependent Sorts 116
9.5 Exercises 117

10 **Replicating Fama and French Factors** 119
10.1 Data Preparation 119
10.2 Portfolio Sorts 121
10.3 Fama and French Factor Returns 122
10.4 Replication Evaluation 123
10.5 Exercises 125

11 **Fama-MacBeth Regressions** 127
11.1 Data Preparation 128
11.2 Cross-sectional Regression 129
11.3 Time-Series Aggregation 130
11.4 Exercises 131

IV Modeling & Machine Learning **133**

12 **Fixed Effects and Clustered Standard Errors** 135
12.1 Data Preparation 136
12.2 Fixed Effects 138
12.3 Clustering Standard Errors 142
12.4 Exercises 144

13 **Difference in Differences** 145
13.1 Data Preparation 146
13.2 Panel Regressions 148
13.3 Visualizing Parallel Trends 150
13.4 Exercises 155

14 **Factor Selection via Machine Learning** 157
14.1 Brief Theoretical Background 158
14.1.1 Ridge regression 159
14.1.2 Lasso 160
14.1.3 Elastic Net 160
14.2 Data Preparation 161
14.3 The Tidymodels Workflow 164
14.3.1 Pre-process data 164
14.3.2 Build a model 166
14.3.3 Fit a model 168
14.3.4 Tune a model 171
14.3.5 Parallelized workflow 175
14.4 Exercises 178

15 **Option Pricing via Machine Learning** 179
15.1 Regression Trees and Random Forests 180
15.2 Neural Networks 181
15.3 Option Pricing 182

15.4 Learning Black-Scholes 182
15.4.1 Data simulation 183
15.4.2 Single layer networks and random forests 184
15.4.3 Deep neural networks 185
15.4.4 Universal approximation 187
15.5 Prediction Evaluation 187
15.6 Exercises 189

V Portfolio Optimization **191**

16 Parametric Portfolio Policies **193**
16.1 Data Preparation 193
16.2 Parametric Portfolio Policies 194
16.3 Computing Portfolio Weights 196
16.4 Portfolio Performance 198
16.5 Optimal Parameter Choice 200
16.6 More Model Specifications 202
16.7 Exercises 204

17 Constrained Optimization and Backtesting **205**
17.1 Data Preparation 205
17.2 Recap of Portfolio Choice 206
17.3 Estimation Uncertainty and Transaction Costs 207
17.4 Optimal Portfolio Choice 208
17.5 Constrained Optimization 211
17.6 Out-of-Sample Backtesting 217
17.7 Exercises 220

A Cover Design **223**

B Clean Enhanced TRACE with R **227**

Bibliography **235**

Index **247**

Preface

Why Does This Book Exist?

Financial economics is a vibrant area of research, a central part of all business activities, and at least implicitly relevant to our everyday life. Despite its relevance for our society and a vast number of empirical studies of financial phenomena, one quickly learns that the actual implementation of models to solve problems in the area of financial economics is typically rather opaque. As graduate students, we were particularly surprised by the lack of public code for seminal papers or even textbooks on key concepts of financial economics. The lack of transparent code not only leads to numerous replication efforts (and their failures) but also constitutes a waste of resources on problems that countless others have already solved in secrecy.

This book aims to lift the curtain on reproducible finance by providing a fully transparent code base for many common financial applications. We hope to inspire others to share their code publicly and take part in our journey toward more reproducible research in the future.

Who Should Read This Book?

We write this book for three audiences:

- Students who want to acquire the basic tools required to conduct financial research ranging from undergrad to graduate level. The book's structure is simple enough such that the material is sufficient for self-study purposes.
- Instructors who look for materials to teach courses in empirical finance or financial economics. We provide plenty of examples and focus on intuitive explanations that can easily be adjusted or expanded. At the end of each chapter, we provide exercises that we hope inspire students to dig deeper.
- Data analysts or statisticians who work on issues dealing with financial data and who need practical tools to succeed.

What Will You Learn?

The book is currently divided into five parts:

- Chapter 1 introduces you to important concepts around which our approach to Tidy Finance revolves.
- Chapters 2–4 provide tools to organize your data and prepare the most common data sets used in financial research. Although many important data are behind paywalls, we start by describing different open-source data and how to download them. We then move on to prepare two of the most popular datasets in financial research: CRSP and Compustat. Then, we cover corporate bond data from TRACE. We reuse the data from these chapters in all subsequent chapters. Chapter 5 contains an overview of common alternative data providers for which direct access vie R packages exist.
- Chapters 6–11 deal with key concepts of empirical asset pricing, such as beta estimation, portfolio sorts, performance analysis, and asset pricing regressions.
- Chapters 12–15 apply linear models to panel data and machine learning methods to problems in factor selection and option pricing.
- Chapters 16–17 provide approaches for parametric, constrained portfolio optimization, and backtesting procedures.
  Each chapter is self-contained and can be read individually. Yet the data chapters provide an important background necessary for data management in all other chapters.

What Won't You Learn?

This book is about empirical work. While we assume only basic knowledge of statistics and econometrics, we do not provide detailed treatments of the underlying theoretical models or methods applied in this book. Instead, you find references to the seminal academic work in journal articles or textbooks for more detailed treatments. We believe that our comparative advantage is to provide a thorough implementation of typical approaches such as portfolio sorts, backtesting procedures, regressions, machine learning methods, or other related topics in empirical finance. We enrich our implementations by discussing the needy-greedy choices you face while conducting empirical analyses. We hence refrain from deriving theoretical models or extensively discussing the statistical properties of well-established tools.

Our book is close in spirit to other books that provide fully reproducible code for financial applications. We view them as complementary to our work and want to highlight the differences:

- Regenstein Jr (2018) provides an excellent introduction and discussion of different tools for standard applications in finance (e.g., how to compute returns and sample standard deviations of a time series of stock returns). In contrast, our book clearly focuses on applications of the state-of-the-art for academic research in finance. We thus fill a niche that allows aspiring researchers or instructors to rely on a well-designed code base.
- Coqueret and Guida (2020) constitute a great compendium to our book with respect to applications related to return prediction and portfolio formation. The book primarily targets practitioners and has a hands-on focus. Our book, in contrast, relies on the typical databases used in financial research and focuses on the preparation of such datasets for academic applications. In addition, our chapter on machine learning focuses on factor selection instead of return prediction.

Although we emphasize the importance of reproducible workflow principles, we do not provide introductions to some of the core tools that we relied on to create and maintain this book:

- Version control systems such as Git[1] are vital in managing any programming project. Originally designed to organize the collaboration of software developers, even solo data analysts will benefit from adopting version control. Git also makes it simple to publicly share code and allow others to reproduce your findings. We refer to Bryan (2022) for a gentle introduction to the (sometimes painful) life with Git.
- Good communication of results is a key ingredient to reproducible and transparent research. To compile this book, we heavily draw on a suite of fantastic open-source tools. First, Wickham (2016) provides a highly customizable yet easy-to-use system for creating data visualizations. Wickham and Grolemund (2016) provide an intuitive introduction to creating graphics using this approach. Second, in our daily work and to compile this book, we used the markdown-based authoring framework described in Xie et al. (2018) and Xie et al. (2020). Markdown documents are fully reproducible and support dozens of static and dynamic output formats. Lastly, Xie (2016) tremendously facilitates authoring markdown-based books. We do not provide introductions to these tools, as the resources above already provide easily accessible tutorials.

- Good writing is also important for the presentation of findings. We neither claim to be experts in this domain nor do we try to sound particularly academic. On the contrary, we deliberately use a more colloquial language to describe all the methods and results presented in this book in order to allow our readers to relate more easily to the rather technical content. For those who desire more guidance with respect to formal academic writing for financial economics, we recommend Kiesling (2003), Cochrane (2005), and Jacobsen (2014), who all provide essential tips (condensed to a few pages).

[1]https://git-scm.com/

Why R?

We believe that R (R Core Team, 2022) is among the best choices for a programming language in the area of finance. Some of our favorite features include:

- R is free and open-source, so that you can use it in academic and professional contexts.
- A diverse and active online community works on a broad range of tools.
- A massive set of actively maintained packages for all kinds of applications exists, e.g., data manipulation, visualization, machine learning, etc.
- Powerful tools for communication, e.g., Rmarkdown and shiny, are readily available.
- RStudio is one of the best development environments for interactive data analysis.
- Strong foundations of functional programming are provided.
- Smooth integration with other programming languages, e.g., SQL, Python, C, C++, Fortran, etc.

For more information on why R is great, we refer to Wickham et al. (2019).

Why Tidy?

As you start working with data, you quickly realize that you spend a lot of time reading, cleaning, and transforming your data. In fact, it is often said that more than 80% of data analysis is spent on preparing data. By *tidying data*, we want to structure data sets to facilitate further analyses. As Wickham (2014) puts it:

> [T]idy datasets are all alike, but every messy dataset is messy in its own way. Tidy datasets provide a standardized way to link the structure of a dataset (its physical layout) with its semantics (its meaning).

In its essence, tidy data follows these three principles:

1. Every column is a variable.

2. Every row is an observation.
3. Every cell is a single value.

Throughout this book, we try to follow these principles as best as we can. If you want to learn more about tidy data principles in an informal manner, we refer you to this vignette[2] as part of Wickham and Girlich (2022).

In addition to the data layer, there are also tidy coding principles outlined in the tidy tools manifesto[3] that we try to follow:

1. Reuse existing data structures.
2. Compose simple functions with the pipe.
3. Embrace functional programming.
4. Design for humans.

In particular, we heavily draw on a set of packages called the `tidyverse`[4] (Wickham et al., 2019). The `tidyverse` is a consistent set of packages for all data analysis tasks, ranging from importing and wrangling to visualizing and modeling data with the same grammar. In addition to explicit tidy principles, the `tidyverse` has further benefits: (i) if you master one package, it is easier to master others, and (ii) the core packages are developed and maintained by the Public Benefit Company Posit. These core packages contained in the `tidyverse` are: `ggplot2` (Wickham, 2016), `dplyr` (Wickham et al., 2022a), `tidyr` (Wickham and Girlich, 2022), `readr` (Wickham et al., 2022c), `purrr` (Henry and Wickham, 2020), `tibble` (Müller and Wickham, 2022), `stringr` (Wickham, 2019), and `forcats` (Wickham, 2021).

Throughout the book we use the native pipe `|>`, a powerful tool to clearly express a sequence of operations. Readers familiar with the `tidyverse` may be used to the predecessor `%>%` that is part of the `magrittr` package. For all our applications, the native and `magrittr` pipe behave identically, so we opt for the one that is simpler and part of base R. For a more thorough discussion on the subtle differences between the two pipes, we refer to the second edition[5] of Wickham and Grolemund (2016).

Prerequisites

Before we continue, make sure you have all the software you need for this book:

[2]https://cran.r-project.org/web/packages/tidyr/vignettes/tidy-data.html
[3]https://tidyverse.tidyverse.org/articles/manifesto.html
[4]https://tidyverse.tidyverse.org/index.html
[5]https://r4ds.hadley.nz/workflow-pipes.html

- Install R and RStudio.[6] To get a walk-through of the installation for every major operating system, follow the steps outlined in this summary.[7] The whole process should be done in a few clicks. If you wonder about the difference: R is an open-source language and environment for statistical computing and graphics, free to download and use. While R runs the computations, RStudio is an integrated development environment that provides an interface by adding many convenient features and tools. We suggest doing all the coding in RStudio.
- Open RStudio and install the `tidyverse`. Not sure how it works? You will find helpful information on how to install packages in this brief summary[8].

If you are new to R, we recommend starting with the following sources:

- A very gentle and good introduction to the workings of R can be found in the form of the weighted dice project[9]. Once you are done setting up R on your machine, try to follow the instructions in this project.
- The main book on the `tidyverse`, Wickham and Grolemund (2016), is available online and for free: R for Data Science[10] explains the majority of the tools we use in our book.
- If you are an instructor searching to effectively teach R and data science methods, we recommend taking a look at the excellent data science toolbox[11] by Mine Cetinkaya-Rundel.[12]
- RStudio provides a range of excellent cheat sheets[13] with extensive information on how to use the `tidyverse` packages.

Colophon

This book was written in RStudio using `bookdown` (Xie, 2016). The website is hosted with GitHub Pages. The complete source is available from GitHub[14]. We generated all plots in this book using `ggplot2` and its classic dark-on-light theme (`theme_bw()`).

This version of the book was built with R (R Core Team, 2022) version 4.2.1 (2022-06-23, Funny-Looking Kid) and the following packages:

[6]https://rstudio-education.github.io/hopr/starting.html#starting
[7]https://rstudio-education.github.io/hopr/starting.html#starthng
[8]https://rstudio-education.github.io/hopr/packages2.html
[9]https://rstudio-education.github.io/hopr/project-1-weighted-dice.html
[10]https://r4ds.had.co.nz/introduction.html
[11]https://datasciencebox.org/
[12]https://mine-cr.com/about/
[13]https://www.rstudio.com/resources/cheatsheets/
[14]www.github.com/voigtstefan/tidy_finance

Package	Version
alabama	2022.4-1
bookdown	0.29
broom	1.0.1
dbplyr	2.2.1
devtools	2.4.4
dplyr	1.0.10
fixest	0.10.4
forcats	0.5.2
frenchdata	0.2.0
furrr	0.3.1
ggplot2	3.3.6
glmnet	4.1-4
googledrive	2.0.0
hardhat	1.2.0
jsonlite	1.8.0
kableExtra	1.3.4
keras	2.9.0
knitr	1.40
lmtest	0.9-40
lubridate	1.8.0
purrr	0.3.4
quadprog	1.5-8
ranger	0.14.1
readr	2.1.2
readxl	1.4.1
renv	0.15.5
rlang	1.0.6
rmarkdown	2.16
RPostgres	1.4.4
RSQLite	2.2.17
sandwich	3.0-2
scales	1.2.1
slider	0.2.2
stringr	1.4.1
tibble	3.1.8
tidymodels	1.0.0
tidyquant	1.0.5
tidyr	1.2.1
tidyverse	1.3.2
timetk	2.8.1

Author biographies

Christoph Scheuch is the Director of Product at the social trading platform wikifolio.com. He is responsible for product planning, execution, and monitoring and manages a team of data scientists to analyze user behavior and develop data-driven products. Christoph is also an external lecturer at the Vienna University of Economics and Business, where he teaches finance students how to manage empirical projects.

Stefan Voigt is Assistant Professor of Finance at the Department of Economics at the University of Copenhagen and a research fellow at the Danish Finance Institute. His research focuses on blockchain technology, high-frequency trading, and financial econometrics. Stefan's research has been published in the leading finance and econometrics journals. He received the Danish Finance Institute teaching award 2022 for his courses for students and practitioners on empirical finance based on this book.

Patrick Weiss is a postdoctoral researcher at the Vienna University of Economics and Business and an external lecturer at Reykjavík University. His research activity centers around the intersection of empirical asset pricing and corporate finance. Patrick is especially passionate about empirical asset pricing and has published research in a top journal in financial economics.

Part I

Getting Started

1

Introduction to Tidy Finance

The main aim of this chapter is to familiarize yourself with the tidyverse. We start by downloading and visualizing stock data from Yahoo!Finance. Then we move to a simple portfolio choice problem and construct the efficient frontier. These examples introduce you to our approach of *Tidy Finance*.

1.1 Working with Stock Market Data

At the start of each session, we load the required packages. Throughout the entire book, we always use the `tidyverse` (Wickham et al., 2019). In this chapter, we also load the convenient `tidyquant` package (Dancho and Vaughan, 2022a) to download price data. This package provides a convenient wrapper for various quantitative functions compatible with the `tidyverse`.

You typically have to install a package once before you can load it. In case you have not done this yet, call `install.packages("tidyquant")`. If you have trouble using `tidyquant`, check out the corresponding documentation.[1]

```
library(tidyverse)
library(tidyquant)
```

We first download daily prices for one stock market ticker, e.g., the Apple stock, *AAPL*, directly from the data provider Yahoo!Finance. To download the data, you can use the command `tq_get`. If you do not know how to use it, make sure you read the help file by calling `?tq_get`. We especially recommend taking a look at the examples section of the documentation. We request daily data for a period of more than 20 years.

```
prices <- tq_get("AAPL",
  get = "stock.prices",
  from = "2000-01-01",
```

[1]https://cran.r-project.org/web/packages/tidyquant/vignettes/TQ00-introduction-to-tidyquant.html

DOI: 10.1201/b23237-1

```
  to = "2021-12-31"
)
prices
```

```
# A tibble: 5,535 x 8
  symbol date        open  high   low close    volume adjusted
  <chr>  <date>     <dbl> <dbl> <dbl> <dbl>     <dbl>    <dbl>
1 AAPL   2000-01-03 0.936 1.00  0.908 0.999 535796800    0.853
2 AAPL   2000-01-04 0.967 0.988 0.903 0.915 512377600    0.781
3 AAPL   2000-01-05 0.926 0.987 0.920 0.929 778321600    0.793
4 AAPL   2000-01-06 0.948 0.955 0.848 0.848 767972800    0.724
5 AAPL   2000-01-07 0.862 0.902 0.853 0.888 460734400    0.759
# ... with 5,530 more rows
```

`tq_get` downloads stock market data from Yahoo!Finance if you do not specify another data source. The function returns a tibble with eight quite self-explanatory columns: `symbol`, `date`, the market prices at the `open`, `high`, `low`, and `close`, the daily `volume` (in the number of traded shares), and the `adjusted` price in USD. The adjusted prices are corrected for anything that might affect the stock price after the market closes, e.g., stock splits and dividends. These actions affect the quoted prices, but they have no direct impact on the investors who hold the stock. Therefore, we often rely on adjusted prices when it comes to analyzing the returns an investor would have earned by holding the stock continuously.

Next, we use the `ggplot2` package (Wickham, 2016) to visualize the time series of adjusted prices in Figure 1.1. This package takes care of visualization tasks based on the principles of the grammar of graphics (Wilkinson, 2012).

```
prices |>
  ggplot(aes(x = date, y = adjusted)) +
  geom_line() +
  labs(
    x = NULL,
    y = NULL,
    title = "Apple stock prices between beginning of 2000 and end of 2021"
  )
```

Instead of analyzing prices, we compute daily net returns defined as $r_t = p_t/p_{t-1} - 1$, where p_t is the adjusted day t price. In that context, the function `lag()` is helpful, which returns the previous value in a vector.

```
returns <- prices |>
  arrange(date) |>
  mutate(ret = adjusted / lag(adjusted) - 1) |>
```

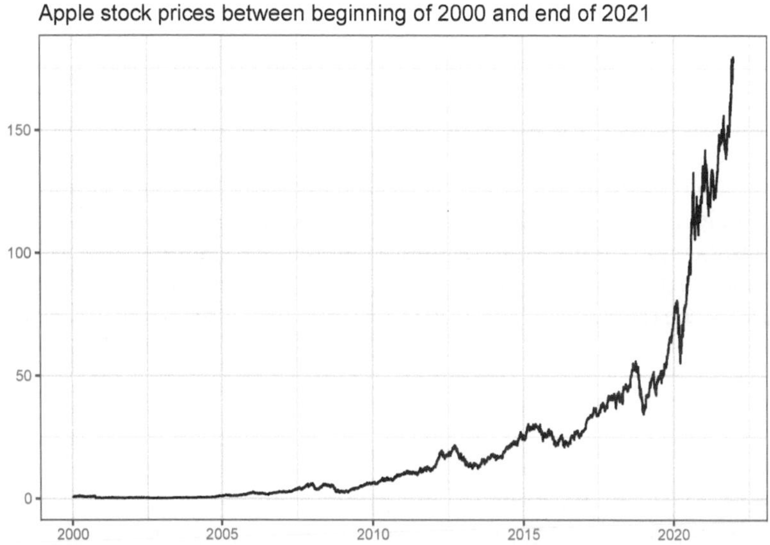

FIGURE 1.1
Prices are in USD, adjusted for dividend payments and stock splits.

```
  select(symbol, date, ret)
returns
```

```
# A tibble: 5,535 x 3
  symbol date           ret
  <chr>  <date>       <dbl>
1 AAPL   2000-01-03 NA
2 AAPL   2000-01-04 -0.0843
3 AAPL   2000-01-05  0.0146
4 AAPL   2000-01-06 -0.0865
5 AAPL   2000-01-07  0.0474
# ... with 5,530 more rows
```

The resulting tibble contains three columns, where the last contains the daily returns (`ret`). Note that the first entry naturally contains a missing value (`NA`) because there is no previous price. Obviously, the use of `lag()` would be meaningless if the time series is not ordered by ascending dates. The command `arrange()` provides a convenient way to order observations in the correct way for our application. In case you want to order observations by descending dates, you can use `arrange(desc(date))`.

For the upcoming examples, we remove missing values as these would require separate treatment when computing, e.g., sample averages. In general, however, make sure you understand why `NA` values occur and carefully examine if you can simply get rid of these observations.

```
returns <- returns |>
  drop_na(ret)
```

Next, we visualize the distribution of daily returns in a histogram in Figure 1.2. For convenience, we multiply the returns by 100 to get returns in percent for the visualizations. Additionally, we add a dashed line that indicates the 5 percent quantile of the daily returns to the histogram, which is a (crude) proxy for the worst return of the stock with a probability of at most 5 percent. The 5 percent quantile is closely connected to the (historical) value-at-risk, a risk measure commonly monitored by regulators. We refer to Tsay (2010) for a more thorough introduction to stylized facts of returns.

```
quantile_05 <- quantile(returns |> pull(ret) * 100, probs = 0.05)

returns |>
  ggplot(aes(x = ret * 100)) +
  geom_histogram(bins = 100) +
  geom_vline(aes(xintercept = quantile_05),
    linetype = "dashed"
  ) +
  labs(
    x = NULL,
    y = NULL,
    title = "Distribution of daily Apple stock returns in percent"
  )
```

Here, `bins = 100` determines the number of bins used in the illustration and hence implicitly the width of the bins. Before proceeding, make sure you understand how to use the geom `geom_vline()` to add a dashed line that indicates the 5 percent quantile of the daily returns. A typical task before proceeding with *any* data is to compute summary statistics for the main variables of interest.

```
returns |>
  mutate(ret = ret * 100) |>
  summarize(across(
    ret,
    list(
      daily_mean = mean,
      daily_sd = sd,
      daily_min = min,
      daily_max = max
    )
  ))
```

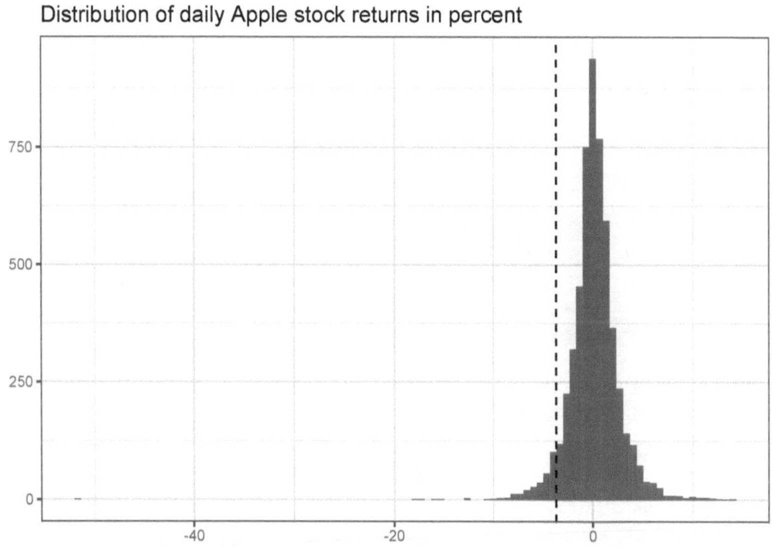

FIGURE 1.2
The dotted vertical line indicates the historical 5 percent quantile.

```
# A tibble: 1 x 4
  ret_daily_mean ret_daily_sd ret_daily_min ret_daily_max
           <dbl>        <dbl>         <dbl>         <dbl>
1          0.130         2.52         -51.9          13.9
```

We see that the maximum *daily* return was 13.905 percent. Perhaps not surprisingly, the average daily return is close to but slightly above 0. In line with the illustration above, the large losses on the day with the minimum returns indicate a strong asymmetry in the distribution of returns.

You can also compute these summary statistics for each year individually by imposing `group_by(year = year(date))`, where the call `year(date)` returns the year. More specifically, the few lines of code below compute the summary statistics from above for individual groups of data defined by year. The summary statistics, therefore, allow an eyeball analysis of the time-series dynamics of the return distribution.

```
returns |>
  mutate(ret = ret * 100) |>
  group_by(year = year(date)) |>
  summarize(across(
    ret,
    list(
      daily_mean = mean,
      daily_sd = sd,
      daily_min = min,
      daily_max = max
```

```
    ),
    .names = "{.fn}"
  )) |>
  print(n = Inf)
```

```
# A tibble: 22 x 5
    year daily_mean daily_sd daily_min daily_max
   <dbl>      <dbl>    <dbl>     <dbl>     <dbl>
 1  2000   -0.346       5.49    -51.9      13.7
 2  2001    0.233       3.93    -17.2      12.9
 3  2002   -0.121       3.05    -15.0       8.46
 4  2003    0.186       2.34     -8.14     11.3
 5  2004    0.470       2.55     -5.58     13.2
 6  2005    0.349       2.45     -9.21      9.12
 7  2006    0.0949      2.43     -6.33     11.8
 8  2007    0.366       2.38     -7.02     10.5
 9  2008   -0.265       3.67    -17.9      13.9
10  2009    0.382       2.14     -5.02      6.76
11  2010    0.183       1.69     -4.96      7.69
12  2011    0.104       1.65     -5.59      5.89
13  2012    0.130       1.86     -6.44      8.87
14  2013    0.0472      1.80    -12.4       5.14
15  2014    0.145       1.36     -7.99      8.20
16  2015    0.00199     1.68     -6.12      5.74
17  2016    0.0575      1.47     -6.57      6.50
18  2017    0.164       1.11     -3.88      6.10
19  2018   -0.00573     1.81     -6.63      7.04
20  2019    0.266       1.65     -9.96      6.83
21  2020    0.281       2.94    -12.9      12.0
22  2021    0.133       1.58     -4.17      5.39
```

In case you wonder: the additional argument `.names = "{.fn}"` in `across()` determines how to name the output columns. The specification is rather flexible and allows almost arbitrary column names, which can be useful for reporting. The `print()` function simply controls the output options for the R console.

1.2 Scaling Up the Analysis

As a next step, we generalize the code from before such that all the computations can handle an arbitrary vector of tickers (e.g., all constituents of an index). Following tidy principles, it is quite easy to download the data, plot the price time series, and tabulate the summary statistics for an arbitrary number of assets.

This is where the `tidyverse` magic starts: tidy data makes it extremely easy to generalize the computations from before to as many assets as you like. The following code takes any vector of tickers, e.g., `ticker <- c("AAPL", "MMM", "BA")`, and automates the download as well as the plot of the price time series. In the end, we create the table of summary statistics for an arbitrary number of assets. We perform the analysis with data from all current constituents of the Dow Jones Industrial Average index.[2]

```
ticker <- tq_index("DOW")
ticker
```

```
# A tibble: 30 x 8
  symbol company            ident~1 sedol weight sector share~2 local~3
  <chr>  <chr>              <chr>   <chr>  <dbl> <chr>    <dbl> <chr>
1 UNH    UnitedHealth Gr~ 91324P~ 2917~ 0.115  Healt~ 5715189 USD
2 GS     Goldman Sachs G~ 38141G~ 2407~ 0.0659 Finan~ 5715189 USD
3 HD     Home Depot Inc.  437076~ 2434~ 0.0608 Consu~ 5715189 USD
4 MCD    McDonald's Corp~ 580135~ 2550~ 0.0535 Consu~ 5715189 USD
5 MSFT   Microsoft Corpo~ 594918~ 2588~ 0.0535 Infor~ 5715189 USD
# ... with 25 more rows, and abbreviated variable names
#   1: identifier, 2: shares_held, 3: local_currency
```

Conveniently, `tidyquant` provides a function to get all stocks in a stock index with a single call (similarly, `tq_exchange("NASDAQ")` delivers all stocks currently listed on the NASDAQ exchange).

```
index_prices <- tq_get(ticker,
  get = "stock.prices",
  from = "2000-01-01",
  to = "2022-12-31"
)
```

The resulting tibble contains 163613 daily observations for 30 different corporations. Figure 1.3 illustrates the time series of downloaded *adjusted* prices for each of the constituents of the Dow Jones index. Make sure you understand every single line of code! (What are the arguments of `aes()`? Which alternative `geoms` could you use to visualize the time series? Hint: if you do not know the answers try to change the code to see what difference your intervention causes.)

[2]https://en.wikipedia.org/wiki/Dow_Jones_Industrial_Average

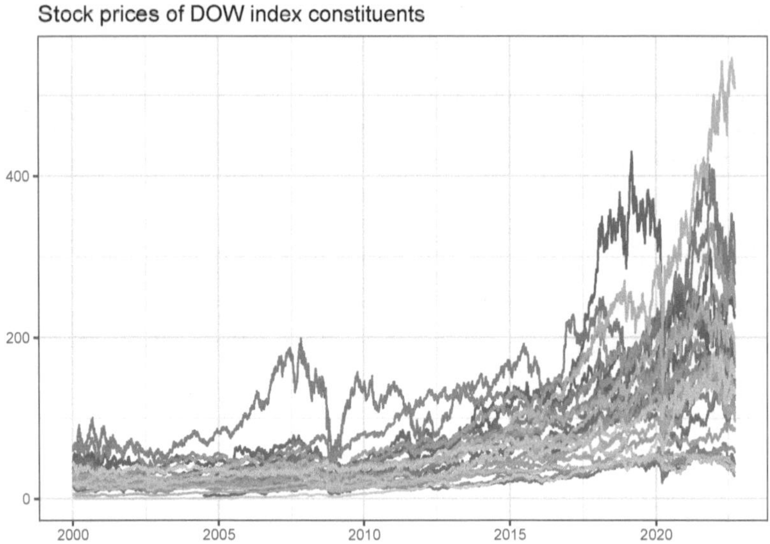

FIGURE 1.3
Prices in USD, adjusted for dividend payments and stock splits.

```
index_prices |>
  ggplot(aes(
    x = date,
    y = adjusted,
    color = symbol
  )) +
  geom_line() +
  labs(
    x = NULL,
    y = NULL,
    color = NULL,
    title = "Stock prices of DOW index constituents"
  ) +
  theme(legend.position = "none")
```

Do you notice the small differences relative to the code we used before? `tq_get(ticker)` returns a tibble for several symbols as well. All we need to do to illustrate all tickers simultaneously is to include `color = symbol` in the `ggplot2` aesthetics. In this way, we generate a separate line for each ticker. Of course, there are simply too many lines on this graph to identify the individual stocks properly, but it illustrates the point well.

The same holds for stock returns. Before computing the returns, we use `group_by(symbol)` such that the `mutate()` command is performed for each symbol individually. The same logic also applies to the computation of summary statistics: `group_by(symbol)` is the key to aggregating the time series into ticker-specific variables of interest.

```
all_returns <- index_prices |>
  group_by(symbol) |>
  mutate(ret = adjusted / lag(adjusted) - 1) |>
  select(symbol, date, ret) |>
  drop_na(ret)

all_returns |>
  mutate(ret = ret * 100) |>
  group_by(symbol) |>
  summarize(across(
    ret,
    list(
      daily_mean = mean,
      daily_sd = sd,
      daily_min = min,
      daily_max = max
    ),
    .names = "{.fn}"
  )) |>
  print(n = Inf)
```

```
# A tibble: 30 x 5
   symbol daily_mean daily_sd daily_min daily_max
   <chr>       <dbl>    <dbl>     <dbl>     <dbl>
 1 AMGN       0.0466     1.97     -13.4      15.1
 2 AXP        0.0508     2.30     -17.6      21.9
 3 BA         0.0528     2.23     -23.8      24.3
 4 CAT        0.0645     2.04     -14.5      14.7
 5 CRM        0.113      2.69     -27.1      26.0
 6 CSCO       0.0289     2.38     -16.2      24.4
 7 CVX        0.0514     1.76     -22.1      22.7
 8 DIS        0.0435     1.94     -18.4      16.0
 9 DOW        0.0414     2.63     -21.7      20.9
10 GS         0.0525     2.32     -19.0      26.5
11 HD         0.0517     1.94     -28.7      14.1
12 HON        0.0479     1.94     -17.4      28.2
13 IBM        0.0247     1.65     -15.5      12.0
14 INTC       0.0285     2.36     -22.0      20.1
15 JNJ        0.0399     1.22     -15.8      12.2
16 JPM        0.0544     2.43     -20.7      25.1
17 KO         0.0318     1.32     -10.1      13.9
```

```
18 MCD      0.0519   1.48   -15.9   18.1
19 MMM      0.0368   1.50   -12.9   12.6
20 MRK      0.0340   1.68   -26.8   13.0
21 MSFT     0.0513   1.93   -15.6   19.6
22 NKE      0.0711   1.92   -19.8   15.5
23 PG       0.0355   1.34   -30.2   12.0
24 TRV      0.0536   1.84   -20.8   25.6
25 UNH      0.0986   1.98   -18.6   34.8
26 V        0.0900   1.90   -13.6   15.0
27 VZ       0.0236   1.51   -11.8   14.6
28 WBA      0.0258   1.81   -15.0   16.6
29 WMT      0.0302   1.50   -11.4   11.7
30 AAPL     0.124    2.51   -51.9   13.9
```

Note that you are now also equipped with all tools to download price data for *each* ticker listed in the S&P 500 index with the same number of lines of code. Just use `ticker <- tq_index("SP500")`, which provides you with a tibble that contains each symbol that is (currently) part of the S&P 500. However, don't try this if you are not prepared to wait for a couple of minutes because this is quite some data to download!

1.3 Other Forms of Data Aggregation

Of course, aggregation across variables other than `symbol` can also make sense. For instance, suppose you are interested in answering the question: are days with high aggregate trading volume likely followed by days with high aggregate trading volume? To provide some initial analysis on this question, we take the downloaded data and compute aggregate daily trading volume for all Dow Jones constituents in USD. Recall that the column `volume` is denoted in the number of traded shares. Thus, we multiply the trading volume with the daily closing price to get a proxy for the aggregate trading volume in USD. Scaling by `1e9` (R can handle scientific notation) denotes daily trading volume in billion USD.

```
volume <- index_prices |>
  group_by(date) |>
  summarize(volume = sum(volume * close / 1e9))

volume |>
  ggplot(aes(x = date, y = volume)) +
  geom_line() +
  labs(
    x = NULL, y = NULL,
```

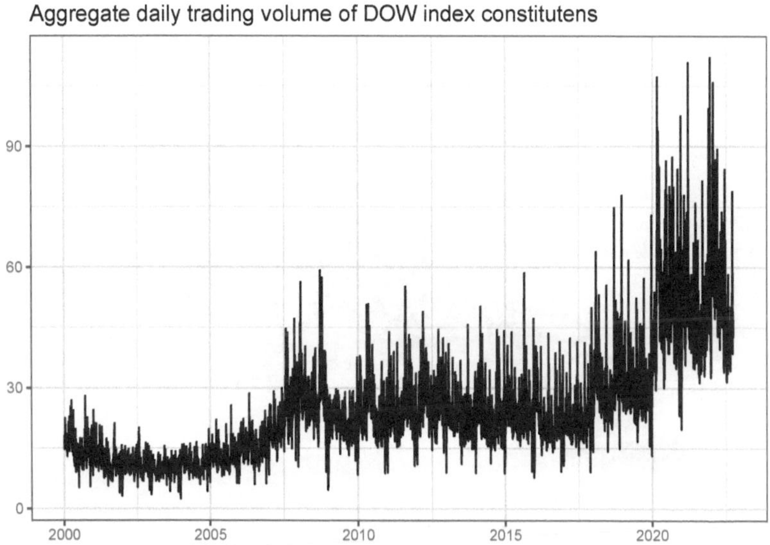

FIGURE 1.4
Total daily trading volume in billion USD.

```
    title = "Aggregate daily trading volume of DOW index constitutens"
  )
```

Figure 1.4 indicates a clear upward trend in aggregated daily trading volume. In particular, since the outbreak of the COVID-19 pandemic, markets have processed substantial trading volumes, as analyzed, for instance, by Goldstein et al. (2021). One way to illustrate the persistence of trading volume would be to plot volume on day t against volume on day $t-1$ as in the example below. In Figure 1.5, we add a dotted 45°-line to indicate a hypothetical one-to-one relation by `geom_abline()`, addressing potential differences in the axes' scales.

```
volume |>
  ggplot(aes(x = lag(volume), y = volume)) +
  geom_point() +
  geom_abline(aes(intercept = 0, slope = 1),
    linetype = "dashed"
  ) +
  labs(
    x = "Previous day aggregate trading volume",
    y = "Aggregate trading volume",
    title = "Persistence in daily trading volume of DOW index constituents"
  )
```

```
Warning: Removed 1 rows containing missing values (geom_point).
```

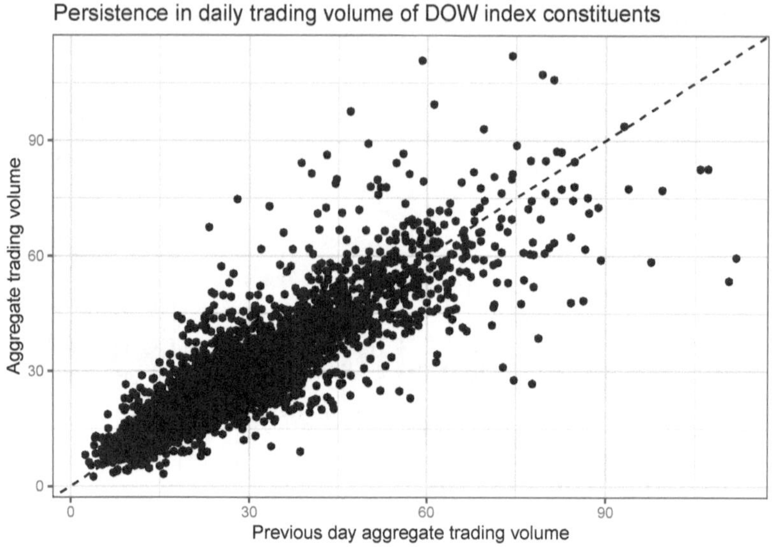

FIGURE 1.5
Total daily trading volume in billion USD.

Do you understand where the warning `## Warning: Removed 1 rows containing missing values (geom_point).` comes from and what it means? Purely eye-balling reveals that days with high trading volume are often followed by similarly high trading volume days.

1.4 Portfolio Choice Problems

In the previous part, we show how to download stock market data and inspect it with graphs and summary statistics. Now, we move to a typical question in Finance: how to allocate wealth across different assets optimally. The standard framework for optimal portfolio selection considers investors that prefer higher future returns but dislike future return volatility (defined as the square root of the return variance): the *mean-variance investor* (Markowitz, 1952).

An essential tool to evaluate portfolios in the mean-variance context is the *efficient frontier*, the set of portfolios which satisfies the condition that no other portfolio exists with a higher expected return but with the same volatility (the square root of the variance, i.e., the risk), see, e.g., Merton (1972). We compute and visualize the efficient frontier for several stocks. First, we extract each asset's *monthly* returns. In order to keep things simple, we work with a balanced panel and exclude DOW constituents for which we do not observe a price on every single trading day since the year 2000.

```
index_prices <- index_prices |>
  group_by(symbol) |>
  mutate(n = n()) |>
  ungroup() |>
  filter(n == max(n)) |>
  select(-n)

returns <- index_prices |>
  mutate(month = floor_date(date, "month")) |>
  group_by(symbol, month) |>
  summarize(price = last(adjusted), .groups = "drop_last") |>
  mutate(ret = price / lag(price) - 1) |>
  drop_na(ret) |>
  select(-price)
```

Here, `floor_date()` is a function from the `lubridate` package (Grolemund and Wickham, 2011), which provides useful functions to work with dates and times.

Next, we transform the returns from a tidy tibble into a $(T \times N)$ matrix with one column for each of the N tickers and one row for each of the T trading days to compute the sample average return vector

$$\hat{\mu} = \frac{1}{T}\sum_{t=1}^{T} r_t$$

where r_t is the N vector of returns on date t and the sample covariance matrix

$$\hat{\Sigma} = \frac{1}{T-1}\sum_{t=1}^{T}(r_t - \hat{\mu})(r_t - \hat{\mu})'.$$

We achieve this by using `pivot_wider()` with the new column names from the column `symbol` and setting the values to `ret`. We compute the vector of sample average returns and the sample variance-covariance matrix, which we consider as proxies for the parameters of the distribution of future stock returns. Thus, for simplicity, we refer to Σ and μ instead of explicitly highlighting that the sample moments are estimates. In later chapters, we discuss the issues that arise once we take estimation uncertainty into account.

```
returns_matrix <- returns |>
  pivot_wider(
    names_from = symbol,
    values_from = ret
```

```
  ) |>
  select(-month)

Sigma <- cov(returns_matrix)
mu <- colMeans(returns_matrix)
```

Then, we compute the minimum variance portfolio weights ω_mvp as well as the expected portfolio return $\omega_\text{mvp}'\mu$ and volatility $\sqrt{\omega_\text{mvp}'\Sigma\omega_\text{mvp}}$ of this portfolio. Recall that the minimum variance portfolio is the vector of portfolio weights that are the solution to

$$\omega_\text{mvp} = \arg\min w'\Sigma w \text{ s.t. } \sum_{i=1}^N w_i = 1.$$

The constraint that weights sum up to one simply implies that all funds are distributed across the available asset universe, i.e., there is no possibility to retain cash. It is easy to show analytically that $\omega_\text{mvp} = \frac{\Sigma^{-1}\iota}{\iota'\Sigma^{-1}\iota}$, where ι is a vector of ones and Σ^{-1} is the inverse of Σ.

```
N <- ncol(returns_matrix)
iota <- rep(1, N)
mvp_weights <- solve(Sigma) %*% iota
mvp_weights <- mvp_weights / sum(mvp_weights)

tibble(
  average_ret = as.numeric(t(mvp_weights) %*% mu),
  volatility = as.numeric(sqrt(t(mvp_weights) %*% Sigma %*% mvp_weights))
)
```

```
# A tibble: 1 x 2
  average_ret volatility
        <dbl>      <dbl>
1     0.00770     0.0315
```

The command `solve(A, b)` returns the solution of a system of equations $Ax = b$. If `b` is not provided, as in the example above, it defaults to the identity matrix such that `solve(Sigma)` delivers Σ^{-1} (if a unique solution exists).
Note that the *monthly* volatility of the minimum variance portfolio is of the same order of magnitude as the *daily* standard deviation of the individual components. Thus, the diversification benefits in terms of risk reduction are tremendous!

Next, we set out to find the weights for a portfolio that achieves, as an example, three times the expected return of the minimum variance portfolio. However, mean-variance investors are not interested in any portfolio that achieves the required return but rather in the efficient portfolio, i.e., the portfolio with the lowest standard deviation. If you wonder where the solution ω_{eff} comes from: The efficient portfolio is chosen by an investor who aims to achieve minimum variance *given a minimum acceptable expected return* $\bar{\mu}$. Hence, their objective function is to choose ω_{eff} as the solution to

$$\omega_{\text{eff}}(\bar{\mu}) = \arg\min w'\Sigma w \text{ s.t. } w'\iota = 1 \text{ and } \omega'\mu \geq \bar{\mu}.$$

The code below implements the analytic solution to this optimization problem for a benchmark return $\bar{\mu}$, which we set to 3 times the expected return of the minimum variance portfolio. We encourage you to verify that it is correct.

```
mu_bar <- 3 * t(mvp_weights) %*% mu

C <- as.numeric(t(iota) %*% solve(Sigma) %*% iota)
D <- as.numeric(t(iota) %*% solve(Sigma) %*% mu)
E <- as.numeric(t(mu) %*% solve(Sigma) %*% mu)

lambda_tilde <- as.numeric(2 * (mu_bar - D / C) / (E - D^2 / C))
efp_weights <- mvp_weights +
  lambda_tilde / 2 * (solve(Sigma) %*% mu - D * mvp_weights)
```

1.5 The Efficient Frontier

The mutual fund separation theorem states that as soon as we have two efficient portfolios (such as the minimum variance portfolio w_{mvp} and the efficient portfolio for a higher required level of expected returns $\omega_{\text{eff}}(\bar{\mu})$, we can characterize the entire efficient frontier by combining these two portfolios. That is, any linear combination of the two portfolio weights will again represent an efficient portfolio. The code below implements the construction of the *efficient frontier*, which characterizes the highest expected return achievable at each level of risk. To understand the code better, make sure to familiarize yourself with the inner workings of the `for` loop.

```
c <- seq(from = -0.4, to = 1.9, by = 0.01)
res <- tibble(
  c = c,
  mu = NA,
  sd = NA
)

for (i in seq_along(c)) {
  w <- (1 - c[i]) * mvp_weights + (c[i]) * efp_weights
  res$mu[i] <- 12 * 100 * t(w) %*% mu
```

```
  res$sd[i] <- 12 * sqrt(100) * sqrt(t(w) %*% Sigma %*% w)
}
```

The code above proceeds in two steps: First, we compute a vector of combination weights c and then we evaluate the resulting linear combination with $c \in \mathbb{R}$:

$$w^* = cw_{\text{eff}}(\bar{\mu}) + (1-c)w_{mvp} = \omega_{\text{mvp}} + \frac{\lambda^*}{2}\left(\Sigma^{-1}\mu - \frac{D}{C}\Sigma^{-1}\iota\right)$$

with $\lambda^* = 2\frac{c\bar{\mu}+(1-c)\tilde{\mu}-D/C}{E-D^2/C}$. where $C = \iota'\Sigma^{-1}\iota$, $D = \iota'\Sigma^{-1}\mu$, and $E = \mu'\Sigma^{-1}\mu$. Finally, it is simple to visualize the efficient frontier alongside the two efficient portfolios within one powerful figure using `ggplot2` (see Figure 1.6). We also add the individual stocks in the same call. We compute annualized returns based on the simple assumption that monthly returns are independent and identically distributed. Thus, the average annualized return is just 12 times the expected monthly return.

```
res |>
  ggplot(aes(x = sd, y = mu)) +
  geom_point() +
  geom_point(
    data = res |> filter(c %in% c(0, 1)),
    size = 4
  ) +
  geom_point(
    data = tibble(
      mu = 12 * 100 * mu,
      sd = 12 * 10 * sqrt(diag(Sigma))
    ),
    aes(y = mu, x = sd), size = 1
  ) +
  labs(
    x = "Annualized standard deviation (in percent)",
    y = "Annualized expected return (in percent)",
    title = "Efficient frontier for DOW index constituents"
  )
```

The line in Figure 1.6 indicates the efficient frontier: the set of portfolios a mean-variance efficient investor would choose from. Compare the performance relative to the individual assets (the dots) – it should become clear that diversifying yields massive performance gains (at least as long as we take the parameters Σ and μ as given).

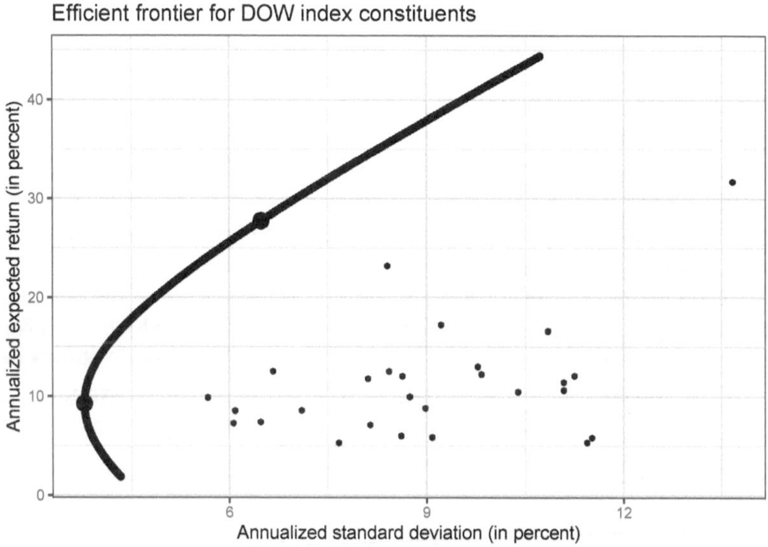

FIGURE 1.6
The big dots indicate the location of the minimum variance and efficient tangency portfolios, respectively. The small dots indicate the location of the individual constituents.

1.6 Exercises

1. Download daily prices for another stock market ticker of your choice from Yahoo!Finance with `tq_get()` from the `tidyquant` package. Plot two time series of the ticker's un-adjusted and adjusted closing prices. Explain the differences.
2. Compute daily net returns for the asset and visualize the distribution of daily returns in a histogram. Also, use `geom_vline()` to add a dashed red line that indicates the 5 percent quantile of the daily returns within the histogram. Compute summary statistics (mean, standard deviation, minimum and maximum) for the daily returns
3. Take your code from before and generalize it such that you can perform all the computations for an arbitrary vector of tickers (e.g., `ticker <- c("AAPL", "MMM", "BA")`). Automate the download, the plot of the price time series, and create a table of return summary statistics for this arbitrary number of assets.
4. Consider the research question: Are days with high aggregate trading volume often also days with large absolute price changes? Find an appropriate visualization to analyze the question.

5. Compute monthly returns from the downloaded stock market prices. Compute the vector of historical average returns and the sample variance-covariance matrix. Compute the minimum variance portfolio weights and the portfolio volatility and average returns. Visualize the mean-variance efficient frontier. Choose one of your assets and identify the portfolio which yields the same historical volatility but achieves the highest possible average return.
6. In the portfolio choice analysis, we restricted our sample to all assets trading every day since 2000. How is such a decision a problem when you want to infer future expected portfolio performance from the results?
7. The efficient frontier characterizes the portfolios with the highest expected return for different levels of risk, i.e., standard deviation. Identify the portfolio with the highest expected return per standard deviation. Hint: the ratio of expected return to standard deviation is an important concept in Finance.

Part II

Financial Data

2

Accessing & Managing Financial Data

In this chapter, we suggest a way to organize your financial data. Everybody, who has experience with data, is also familiar with storing data in various formats like CSV, XLS, XLSX, or other delimited value storage. Reading and saving data can become very cumbersome in the case of using different data formats, both across different projects and across different programming languages. Moreover, storing data in delimited files often leads to problems with respect to column type consistency. For instance, date-type columns frequently lead to inconsistencies across different data formats and programming languages.

This chapter shows how to import different open source data sets. Specifically, our data comes from the application programming interface (API) of Yahoo!Finance, a downloaded standard CSV file, an XLSX file stored in a public Google Drive repository, and other macroeconomic time series. We store all the data in a *single* database, which serves as the only source of data in subsequent chapters. We conclude the chapter by providing some tips on managing databases.

First, we load the global packages that we use throughout this chapter. Later on, we load more packages in the sections where we need them.

```
library(tidyverse)
library(lubridate)
library(scales)
```

The package `lubridate` provides convenient tools to work with dates and times (Grolemund and Wickham, 2011). The package `scales` (Wickham and Seidel, 2022) provides useful scale functions for visualizations.

Moreover, we initially define the date range for which we fetch and store the financial data, making future data updates tractable. In case you need another time frame, you can adjust the dates below. Our data starts with 1960 since most asset pricing studies use data from 1962 on.

```
start_date <- ymd("1960-01-01")
end_date <- ymd("2021-12-31")
```

DOI: 10.1201/b23237-2

2.1 Fama-French Data

We start by downloading some famous Fama-French factors (e.g., Fama and French, 1993) and portfolio returns commonly used in empirical asset pricing. Fortunately, there is a neat package by Nelson Areal[1] that allows us to access the data easily: the `frenchdata` package provides functions to download and read data sets from Prof. Kenneth French finance data library[2] (Areal, 2021).

```
library(frenchdata)
```

We can use the main function of the package to download monthly Fama-French factors. The set *3 Factors* includes the return time series of the market, size, and value factors alongside the risk-free rates. Note that we have to do some manual work to correctly parse all the columns and scale them appropriately, as the raw Fama-French data comes in a very unpractical data format. For precise descriptions of the variables, we suggest consulting Prof. Kenneth French's finance data library directly. If you are on the site, check the raw data files to appreciate the time you can save thanks to `frenchdata`.

```
factors_ff_monthly_raw <- download_french_data("Fama/French 3 Factors")
factors_ff_monthly <- factors_ff_monthly_raw$subsets$data[[1]] |>
  transmute(
    month = floor_date(ymd(str_c(date, "01")), "month"),
    rf = as.numeric(RF) / 100,
    mkt_excess = as.numeric(`Mkt-RF`) / 100,
    smb = as.numeric(SMB) / 100,
    hml = as.numeric(HML) / 100
  ) |>
  filter(month >= start_date & month <= end_date)
```

It is straightforward to download the corresponding *daily* Fama-French factors with the same function.

```
factors_ff_daily_raw <- download_french_data("Fama/French 3 Factors [Daily]")
factors_ff_daily <- factors_ff_daily_raw$subsets$data[[1]] |>
  transmute(
    date = ymd(date),
    rf = as.numeric(RF) / 100,
    mkt_excess = as.numeric(`Mkt-RF`) / 100,
    smb = as.numeric(SMB) / 100,
```

[1] https://github.com/nareal/frenchdata/

[2] https://mba.tuck.dartmouth.edu/pages/faculty/ken.french/data_library.html

```
    hml = as.numeric(HML) / 100
  ) |>
  filter(date >= start_date & date <= end_date)
```

In a subsequent chapter, we also use the 10 monthly industry portfolios, so let us fetch that data, too.

```
industries_ff_monthly_raw <- download_french_data("10 Industry Portfolios")
industries_ff_monthly <- industries_ff_monthly_raw$subsets$data[[1]] |>
  mutate(month = floor_date(ymd(str_c(date, "01")), "month")) |>
  mutate(across(where(is.numeric), ~ . / 100)) |>
  select(month, everything(), -date) |>
  filter(month >= start_date & month <= end_date)
```

It is worth taking a look at all available portfolio return time series from Kenneth French's homepage. You should check out the other sets by calling `get_french_data_list()`.

2.2 q-Factors

In recent years, the academic discourse experienced the rise of alternative factor models, e.g., in the form of the Hou et al. (2014) *q*-factor model. We refer to the extended background[3] information provided by the original authors for further information. The *q* factors can be downloaded directly from the authors' homepage from within `read_csv()`.

We also need to adjust this data. First, we discard information we will not use in the remainder of the book. Then, we rename the columns with the "R_"-prescript using regular expressions and write all column names in lowercase. You should always try sticking to a consistent style for naming objects, which we try to illustrate here – the emphasis is on *try*. You can check out style guides available online, e.g., Hadley Wickham's `tidyverse` style guide.[4]

```
factors_q_monthly_link <-
  "http://global-q.org/uploads/1/2/2/6/122679606/q5_factors_monthly_2021.csv"

factors_q_monthly <- read_csv(factors_q_monthly_link) |>
  mutate(month = ymd(str_c(year, month, "01", sep = "-"))) |>
```

[3]http://global-q.org/background.html
[4]https://style.tidyverse.org/index.html

```
  select(-R_F, -R_MKT, -year) |>
  rename_with(~ str_remove(., "R_")) |>
  rename_with(~ str_to_lower(.)) |>
  mutate(across(-month, ~ . / 100)) |>
  filter(month >= start_date & month <= end_date)
```

2.3 Macroeconomic Predictors

Our next data source is a set of macroeconomic variables often used as predictors for the equity premium. Welch and Goyal (2008) comprehensively reexamine the performance of variables suggested by the academic literature to be good predictors of the equity premium. The authors host the data updated to 2021 on Amit Goyal's website.[5] Since the data is an XLSX-file stored on a public Google drive location, we need additional packages to access the data directly from our R session. Therefore, we load `readxl` to read the XLSX-file (Wickham and Bryan, 2022) and `googledrive` for the Google drive connection (D'Agostino McGowan and Bryan, 2021).

```
library(readxl)
library(googledrive)
```

Usually, you need to authenticate if you interact with Google drive directly in R. Since the data is stored via a public link, we can proceed without any authentication.

```
drive_deauth()
```

The `drive_download()` function from the `googledrive` package allows us to download the data and store it locally.

```
macro_predictors_link <-
  "https://docs.google.com/spreadsheets/d/1OArfD2Wv9IvGoLkJ8JyoXS0YMQLDZfY2"

drive_download(
  macro_predictors_link,
  path = "data/macro_predictors.xlsx"
)
```

Next, we read in the new data and transform the columns into the variables that we later use:

[5] https://sites.google.com/view/agoyal145

1. The dividend price ratio (dp), the difference between the log of dividends and the log of prices, where dividends are 12-month moving sums of dividends paid on the S&P 500 index, and prices are monthly averages of daily closing prices (Campbell and Shiller, 1988; Campbell and Yogo, 2006).
2. Dividend yield (dy), the difference between the log of dividends and the log of lagged prices (Ball, 1978).
3. Earnings price ratio (ep), the difference between the log of earnings and the log of prices, where earnings are 12-month moving sums of earnings on the S&P 500 index (Campbell and Shiller, 1988).
4. Dividend payout ratio (de), the difference between the log of dividends and the log of earnings (Lamont, 1998).
5. Stock variance (svar), the sum of squared daily returns on the S&P 500 index (Guo, 2006).
6. Book-to-market ratio (bm), the ratio of book value to market value for the Dow Jones Industrial Average (Kothari and Shanken, 1997)
7. Net equity expansion (ntis), the ratio of 12-month moving sums of net issues by NYSE listed stocks divided by the total end-of-year market capitalization of NYSE stocks (Campbell et al., 2008).
8. Treasury bills (tbl), the 3-Month Treasury Bill: Secondary Market Rate from the economic research database at the Federal Reserve Bank at St. Louis (Campbell, 1987).
9. Long-term yield (lty), the long-term government bond yield from Ibbotson's Stocks, Bonds, Bills, and Inflation Yearbook (Welch and Goyal, 2008).
10. Long-term rate of returns (ltr), the long-term government bond returns from Ibbotson's Stocks, Bonds, Bills, and Inflation Yearbook (Welch and Goyal, 2008).
11. Term spread (tms), the difference between the long-term yield on government bonds and the Treasury bill (Campbell, 1987).
12. Default yield spread (dfy), the difference between BAA and AAA-rated corporate bond yields (Fama and French, 1989).
13. Inflation (infl), the Consumer Price Index (All Urban Consumers) from the Bureau of Labor Statistics (Campbell and Vuolteenaho, 2004).

For variable definitions and the required data transformations, you can consult the material on Amit Goyal's website[6].

```
macro_predictors <- read_xlsx(
  "data/macro_predictors.xlsx",
  sheet = "Monthly"
) |>
  mutate(month = ym(yyyymm)) |>
  mutate(across(where(is.character), as.numeric)) |>
```

[6]https://sites.google.com/view/agoyal145

```
  mutate(
    IndexDiv = Index + D12,
    logret = log(IndexDiv) - log(lag(IndexDiv)),
    Rfree = log(Rfree + 1),
    rp_div = lead(logret - Rfree, 1), # Future excess market return
    dp = log(D12) - log(Index), # Dividend Price ratio
    dy = log(D12) - log(lag(Index)), # Dividend yield
    ep = log(E12) - log(Index), # Earnings price ratio
    de = log(D12) - log(E12), # Dividend payout ratio
    tms = lty - tbl, # Term spread
    dfy = BAA - AAA # Default yield spread
  ) |>
  select(month, rp_div, dp, dy, ep, de, svar,
    bm = `b/m`, ntis, tbl, lty, ltr,
    tms, dfy, infl
  ) |>
  filter(month >= start_date & month <= end_date) |>
  drop_na()
```

Finally, after reading in the macro predictors to our memory, we remove the raw data file from our temporary storage.

```
file.remove("data/macro_predictors.xlsx")
```

```
[1] TRUE
```

2.4 Other Macroeconomic Data

The Federal Reserve bank of St. Louis provides the Federal Reserve Economic Data (FRED), an extensive database for macroeconomic data. In total, there are 817,000 US and international time series from 108 different sources. As an illustration, we use the already familiar `tidyquant` package to fetch consumer price index (CPI) data that can be found under the CPIAUCNS[7] key.

```
library(tidyquant)

cpi_monthly <- tq_get("CPIAUCNS",
```

[7]https://fred.stlouisfed.org/series/CPIAUCNS

```
  get = "economic.data",
  from = start_date,
  to = end_date
) |>
  transmute(
    month = floor_date(date, "month"),
    cpi = price / price[month == max(month)]
  )
```

To download other time series, we just have to look it up on the FRED website and extract the corresponding key from the address. For instance, the producer price index for gold ores can be found under the PCU2122212122210[8] key. The `tidyquant` package provides access to around 10,000 time series of the FRED database. If your desired time series is not included, we recommend working with the `fredr` package (Boysel and Vaughan, 2021). Note that you need to get an API key to use its functionality. We refer to the package documentation for details.

2.5 Setting Up a Database

Now that we have downloaded some (freely available) data from the web into the memory of our R session let us set up a database to store that information for future use. We will use the data stored in this database throughout the following chapters, but you could alternatively implement a different strategy and replace the respective code.

There are many ways to set up and organize a database, depending on the use case. For our purpose, the most efficient way is to use an SQLite[9] database, which is the C-language library that implements a small, fast, self-contained, high-reliability, full-featured, SQL database engine. Note that SQL[10] (Structured Query Language) is a standard language for accessing and manipulating databases and heavily inspired the `dplyr` functions. We refer to this tutorial[11] for more information on SQL.

There are two packages that make working with SQLite in R very simple: `RSQLite` (Müller et al., 2022) embeds the SQLite database engine in R, and `dbplyr` (Wickham et al., 2022b) is the database back-end for `dplyr`. These packages allow to set up a database to remotely store tables and use these remote database tables as if they are

[8]https://fred.stlouisfed.org/series/PCU2122212122210
[9]https://www.sqlite.org/index.html
[10]https://en.wikipedia.org/wiki/SQL
[11]https://www.w3schools.com/sql/sql_intro.asp

in-memory data frames by automatically converting `dplyr` into SQL. Check out the `RSQLite`[12] and `dbplyr`[13] vignettes for more information.

```
library(RSQLite)
library(dbplyr)
```

An SQLite database is easily created – the code below is really all there is. You do not need any external software. Note that we use the `extended_types=TRUE` option to enable date types when storing and fetching data. Otherwise, date columns are stored and retrieved as integers. We will use the resulting file `tidy_finance.sqlite` in the subfolder `data` for all subsequent chapters to retrieve our data.

```
tidy_finance <- dbConnect(
  SQLite(),
  "data/tidy_finance.sqlite",
  extended_types = TRUE
)
```

Next, we create a remote table with the monthly Fama-French factor data. We do so with the function `dbWriteTable()`, which copies the data to our SQLite-database.

```
  dbWriteTable(tidy_finance,
    "factors_ff_monthly",
    value = factors_ff_monthly,
    overwrite = TRUE
  )
```

We can use the remote table as an in-memory data frame by building a connection via `tbl()`.

```
factors_ff_monthly_db <- tbl(tidy_finance, "factors_ff_monthly")
```

All `dplyr` calls are evaluated lazily, i.e., the data is not in our R session's memory, and the database does most of the work. You can see that by noticing that the output below does not show the number of rows. In fact, the following code chunk only fetches the top 10 rows from the database for printing.

```
factors_ff_monthly_db |>
  select(month, rf)
```

[12]https://cran.r-project.org/web/packages/RSQLite/vignettes/RSQLite.html
[13]https://db.rstudio.com/databases/sqlite/

```
# Source:   SQL [?? x 2]
# Database: sqlite 3.39.3 [data/tidy_finance.sqlite]
  month          rf
  <date>      <dbl>
1 1960-01-01 0.0033
2 1960-02-01 0.0029
3 1960-03-01 0.0035
4 1960-04-01 0.0019
5 1960-05-01 0.0027
# ... with more rows
```

If we want to have the whole table in memory, we need to `collect()` it. You will see that we regularly load the data into the memory in the next chapters.

```
factors_ff_monthly_db |>
  select(month, rf) |>
  collect()
```

```
# A tibble: 744 x 2
  month          rf
  <date>      <dbl>
1 1960-01-01 0.0033
2 1960-02-01 0.0029
3 1960-03-01 0.0035
4 1960-04-01 0.0019
5 1960-05-01 0.0027
# ... with 739 more rows
```

The last couple of code chunks is really all there is to organizing a simple database! You can also share the SQLite database across devices and programming languages.

Before we move on to the next data source, let us also store the other five tables in our new SQLite database.

```
dbWriteTable(tidy_finance,
  "factors_ff_daily",
  value = factors_ff_daily,
  overwrite = TRUE
)

dbWriteTable(tidy_finance,
  "industries_ff_monthly",
  value = industries_ff_monthly,
  overwrite = TRUE
)
```

```
dbWriteTable(tidy_finance,
  "factors_q_monthly",
  value = factors_q_monthly,
  overwrite = TRUE
)

dbWriteTable(tidy_finance,
  "macro_predictors",
  value = macro_predictors,
  overwrite = TRUE
)

dbWriteTable(tidy_finance,
  "cpi_monthly",
  value = cpi_monthly,
  overwrite = TRUE
)
```

From now on, all you need to do to access data that is stored in the database is to follow three steps: (i) Establish the connection to the SQLite database, (ii) call the table you want to extract, and (iii) collect the data. For your convenience, the following steps show all you need in a compact fashion.

```
library(tidyverse)
library(RSQLite)

tidy_finance <- dbConnect(
  SQLite(),
  "data/tidy_finance.sqlite",
  extended_types = TRUE
)

factors_q_monthly <- tbl(tidy_finance, "factors_q_monthly")
factors_q_monthly <- factors_q_monthly |> collect()
```

2.6 Managing SQLite Databases

Finally, at the end of our data chapter, we revisit the SQLite database itself. When you drop database objects such as tables or delete data from tables, the database file size remains unchanged because SQLite just marks the deleted objects as free and reserves their space for future uses. As a result, the database file always grows in size.

To optimize the database file, you can run the `VACUUM` command in the database, which rebuilds the database and frees up unused space. You can execute the command in the database using the `dbSendQuery()` function.

```
dbSendQuery(tidy_finance, "VACUUM")
```

```
<SQLiteResult>
  SQL  VACUUM
  ROWS Fetched: 0 [complete]
       Changed: 0
```

The `VACUUM` command actually performs a couple of additional cleaning steps, which you can read up in this tutorial.[14]

Apart from cleaning up, you might be interested in listing all the tables that are currently in your database. You can do this via the `dbListTables()` function.

```
dbListTables(tidy_finance)
```

```
Warning: Closing open result set, pending rows
```

```
 [1] "beta"                   "compustat"
 [3] "cpi_monthly"            "crsp_daily"
 [5] "crsp_monthly"           "factors_ff_daily"
 [7] "factors_ff_monthly"     "factors_q_monthly"
 [9] "industries_ff_monthly" "macro_predictors"
[11] "mergent"                "trace_enhanced"
```

This function comes in handy if you are unsure about the correct naming of the tables in your database.

2.7 Exercises

1. Download the monthly Fama-French factors manually from Ken French's data library[15] and read them in via `read_csv()`. Validate that you get the same data as via the `frenchdata` package.

[14] https://www.sqlitetutorial.net/sqlite-vacuum/

[15] https://mba.tuck.dartmouth.edu/pages/faculty/ken.french/data_library.html

2. Download the Fama-French 5 factors using the `frenchdata` package. Use `get_french_data_list()` to find the corresponding table name. After the successful download and conversion to the column format that we used above, compare the resulting `rf`, `mkt_excess`, `smb`, and `hml` columns to `factors_ff_monthly`. Explain any differences you might find.

3

WRDS, CRSP, and Compustat

This chapter shows how to connect to Wharton Research Data Services (WRDS)[1], a popular provider of financial and economic data for research applications. We use this connection to download the most commonly used data for stock and firm characteristics, CRSP and Compustat. Unfortunately, this data is not freely available, but most students and researchers typically have access to WRDS through their university libraries. Assuming that you have access to WRDS, we show you how to prepare and merge the databases and store them in the `SQLite`-database introduced in the previous chapter. We conclude this chapter by providing some tips for working with the WRDS database.

First, we load the packages that we use throughout this chapter. Later on, we load more packages in the sections where we need them.

```
library(tidyverse)
library(lubridate)
library(scales)
library(RSQLite)
library(dbplyr)
```

We use the same date range as in the previous chapter to ensure consistency.

```
start_date <- ymd("1960-01-01")
end_date <- ymd("2021-12-31")
```

3.1 Accessing WRDS

WRDS is the most widely used source for asset and firm-specific financial data used in academic settings. WRDS is a data platform that provides data validation, flexible delivery options, and access to many different data sources. The data at WRDS is also organized in an SQL database, although they use the PostgreSQL[2] engine. This

[1] https://wrds-www.wharton.upenn.edu/
[2] https://www.postgresql.org/

DOI: 10.1201/b23237-3

database engine is just as easy to handle with R as SQLite. We use the `RPostgres` package to establish a connection to the WRDS database (Wickham et al., 2022e). Note that you could also use the `odbc` package to connect to a PostgreSQL database, but then you need to install the appropriate drivers yourself. `RPostgres` already contains a suitable driver.

```
library(RPostgres)
```

To establish a connection, you use the function `dbConnect()` with the following arguments. Note that you need to replace the `user` and `password` fields with your own credentials. We defined system variables for the purpose of this book because we obviously do not want (and are not allowed) to share our credentials with the rest of the world.

```
wrds <- dbConnect(
  Postgres(),
  host = "wrds-pgdata.wharton.upenn.edu",
  dbname = "wrds",
  port = 9737,
  sslmode = "require",
  user = Sys.getenv("user"),
  password = Sys.getenv("password")
)
```

The remote connection to WRDS is very useful. Yet, the database itself contains many different tables. You can check the WRDS homepage to identify the table's name you are looking for (if you go beyond our exposition). Alternatively, you can also query the data structure with the function `dbSendQuery()`. If you are interested, there is an exercise below that is based on WRDS' tutorial on "Querying WRDS Data using R".[3] Furthermore, the penultimate section of this chapter shows how to investigate the structure of databases.

3.2 Downloading and Preparing CRSP

The Center for Research in Security Prices (CRSP)[4] provides the most widely used data for US stocks. We use the `wrds` connection object that we just created to first access monthly CRSP return data. Actually, we need three tables to get the desired data: (i) the CRSP monthly security file,

[3] https://wrds-www.wharton.upenn.edu/pages/support/programming-wrds/programming-r/querying-wrds-data-r/
[4] https://crsp.org/

```
msf_db <- tbl(wrds, in_schema("crsp", "msf"))
```

(ii) the identifying information,

```
msenames_db <- tbl(wrds, in_schema("crsp", "msenames"))
```

and (iii) the delisting information.

```
msedelist_db <- tbl(wrds, in_schema("crsp", "msedelist"))
```

We use the three remote tables to fetch the data we want to put into our local database. Just as above, the idea is that we let the WRDS database do all the work and just download the data that we actually need. We apply common filters and data selection criteria to narrow down our data of interest: (i) we keep only data in the time windows of interest, (ii) we keep only US-listed stocks as identified via share codes `shrcd` 10 and 11, and (iii) we keep only months within permno-specific start dates `namedt` and end dates `nameendt`. In addition, we add delisting codes and returns. You can read up in the great textbook of Bali et al. (2016) for an extensive discussion on the filters we apply in the code below.

```
crsp_monthly <- msf_db |>
  filter(date >= start_date & date <= end_date) |>
  inner_join(
    msenames_db |>
      filter(shrcd %in% c(10, 11)) |>
      select(permno, exchcd, siccd, namedt, nameendt),
    by = c("permno")
  ) |>
  filter(date >= namedt & date <= nameendt) |>
  mutate(month = floor_date(date, "month")) |>
  left_join(
    msedelist_db |>
      select(permno, dlstdt, dlret, dlstcd) |>
      mutate(month = floor_date(dlstdt, "month")),
    by = c("permno", "month")
  ) |>
  select(
    permno, # Security identifier
    date, # Date of the observation
    month, # Month of the observation
    ret, # Return
    shrout, # Shares outstanding (in thousands)
    altprc, # Last traded price in a month
    exchcd, # Exchange code
```

```
    siccd, # Industry code
    dlret, # Delisting return
    dlstcd # Delisting code
  ) |>
  collect() |>
  mutate(
    month = ymd(month),
    shrout = shrout * 1000
  )
```

Now, we have all the relevant monthly return data in memory and proceed with preparing the data for future analyses. We perform the preparation step at the current stage since we want to avoid executing the same mutations every time we use the data in subsequent chapters.

The first additional variable we create is market capitalization (`mktcap`), which is the product of the number of outstanding shares `shrout` and the last traded price in a month `altprc`. Note that in contrast to returns '`ret`, these two variables are not adjusted ex-post for any corporate actions like stock splits. Moreover, the `altprc` is negative whenever the last traded price does not exist, and CRSP decides to report the mid-quote of the last available order book instead. Hence, we take the absolute value of the market cap. We also keep the market cap in millions of USD just for convenience as we do not want to print huge numbers in our figures and tables. In addition, we set zero market cap to missing as it makes conceptually little sense (i.e., the firm would be bankrupt).

```
crsp_monthly <- crsp_monthly |>
  mutate(
    mktcap = abs(shrout * altprc) / 1000000,
    mktcap = na_if(mktcap, 0)
  )
```

The next variable we frequently use is the one-month *lagged* market capitalization. Lagged market capitalization is typically used to compute value-weighted portfolio returns, as we demonstrate in a later chapter. The most simple and consistent way to add a column with lagged market cap values is to add one month to each observation and then join the information to our monthly CRSP data.

```
mktcap_lag <- crsp_monthly |>
  mutate(month = month %m+% months(1)) |>
  select(permno, month, mktcap_lag = mktcap)

crsp_monthly <- crsp_monthly |>
  left_join(mktcap_lag, by = c("permno", "month"))
```

If you wonder why we do not use the `lag()` function, e.g., via `crsp_monthly |> group_by(permno) |> mutate(mktcap_lag = lag(mktcap))`, take a look at the exercises.

Next, we follow Bali et al. (2016) in transforming listing exchange codes to explicit exchange names.

```
crsp_monthly <- crsp_monthly |>
  mutate(exchange = case_when(
    exchcd %in% c(1, 31) ~ "NYSE",
    exchcd %in% c(2, 32) ~ "AMEX",
    exchcd %in% c(3, 33) ~ "NASDAQ",
    TRUE ~ "Other"
  ))
```

Similarly, we transform industry codes to industry descriptions following Bali et al. (2016). Notice that there are also other categorizations of industries (e.g., Fama and French, 1997) that are commonly used.

```
crsp_monthly <- crsp_monthly |>
  mutate(industry = case_when(
    siccd >= 1 & siccd <= 999 ~ "Agriculture",
    siccd >= 1000 & siccd <= 1499 ~ "Mining",
    siccd >= 1500 & siccd <= 1799 ~ "Construction",
    siccd >= 2000 & siccd <= 3999 ~ "Manufacturing",
    siccd >= 4000 & siccd <= 4899 ~ "Transportation",
    siccd >= 4900 & siccd <= 4999 ~ "Utilities",
    siccd >= 5000 & siccd <= 5199 ~ "Wholesale",
    siccd >= 5200 & siccd <= 5999 ~ "Retail",
    siccd >= 6000 & siccd <= 6799 ~ "Finance",
    siccd >= 7000 & siccd <= 8999 ~ "Services",
    siccd >= 9000 & siccd <= 9999 ~ "Public",
    TRUE ~ "Missing"
  ))
```

We also construct returns adjusted for delistings as described by Bali et al. (2016). The delisting of a security usually results when a company ceases operations, declares bankruptcy, merges, does not meet listing requirements, or seeks to become private. The adjustment tries to reflect the returns of investors who bought the stock in the month before the delisting and held it until the delisting date. After this transformation, we can drop the delisting returns and codes.

```
crsp_monthly <- crsp_monthly |>
  mutate(ret_adj = case_when(
    is.na(dlstcd) ~ ret,
    !is.na(dlstcd) & !is.na(dlret) ~ dlret,
```

```
    dlstcd %in% c(500, 520, 580, 584) |
      (dlstcd >= 551 & dlstcd <= 574) ~ -0.30,
    dlstcd == 100 ~ ret,
    TRUE ~ -1
  )) |>
  select(-c(dlret, dlstcd))
```

Next, we compute excess returns by subtracting the monthly risk-free rate provided by our Fama-French data. As we base all our analyses on the excess returns, we can drop adjusted returns and the risk-free rate from our tibble. Note that we ensure excess returns are bounded by −1 from below as a return less than −100% makes no sense conceptually. Before we can adjust the returns, we have to connect to our database and load the tibble `factors_ff_monthly`.

```
tidy_finance <- dbConnect(
  SQLite(),
  "data/tidy_finance.sqlite",
  extended_types = TRUE
)

factors_ff_monthly <- tbl(tidy_finance, "factors_ff_monthly") |>
  collect()

crsp_monthly <- crsp_monthly |>
  left_join(factors_ff_monthly |> select(month, rf),
    by = "month"
  ) |>
  mutate(
    ret_excess = ret_adj - rf,
    ret_excess = pmax(ret_excess, -1)
  ) |>
  select(-ret_adj, -rf)
```

Since excess returns and market capitalization are crucial for all our analyses, we can safely exclude all observations with missing returns or market capitalization.

```
crsp_monthly <- crsp_monthly |>
  drop_na(ret_excess, mktcap, mktcap_lag)
```

Finally, we store the monthly CRSP file in our database.

```
  dbWriteTable(tidy_finance,
    "crsp_monthly",
    value = crsp_monthly,
```

```
  overwrite = TRUE
)
```

3.3 First Glimpse of the CRSP Sample

Before we move on to other data sources, let us look at some descriptive statistics of the CRSP sample, which is our main source for stock returns.

Figure 3.1 shows the monthly number of securities by listing exchange over time. NYSE has the longest history in the data, but NASDAQ lists a considerably large number of stocks. The number of stocks listed on AMEX decreased steadily over the last couple of decades. By the end of 2021, there were 2,779 stocks with a primary listing on NASDAQ, 1,395 on NYSE, 145 on AMEX, and only one belonged to the other category.

```
crsp_monthly |>
  count(exchange, date) |>
  ggplot(aes(x = date, y = n, color = exchange, linetype = exchange)) +
  geom_line() +
  labs(
    x = NULL, y = NULL, color = NULL, linetype = NULL,
    title = "Monthly number of securities by listing exchange"
  ) +
  scale_x_date(date_breaks = "10 years", date_labels = "%Y") +
  scale_y_continuous(labels = comma)
```

Next, we look at the aggregate market capitalization grouped by the respective listing exchanges in Figure 3.2. To ensure that we look at meaningful data which is comparable over time, we adjust the nominal values for inflation. In fact, we can use the tables that are already in our database to calculate aggregate market caps by listing exchange and plotting it just as if they were in memory. All values in Figure 3.2 are at the end of 2021 USD to ensure intertemporal comparability. NYSE-listed stocks have by far the largest market capitalization, followed by NASDAQ-listed stocks.

```
tbl(tidy_finance, "crsp_monthly") |>
  left_join(tbl(tidy_finance, "cpi_monthly"), by = "month") |>
  group_by(month, exchange) |>
  summarize(
    mktcap = sum(mktcap, na.rm = TRUE) / cpi,
    .groups = "drop"
```

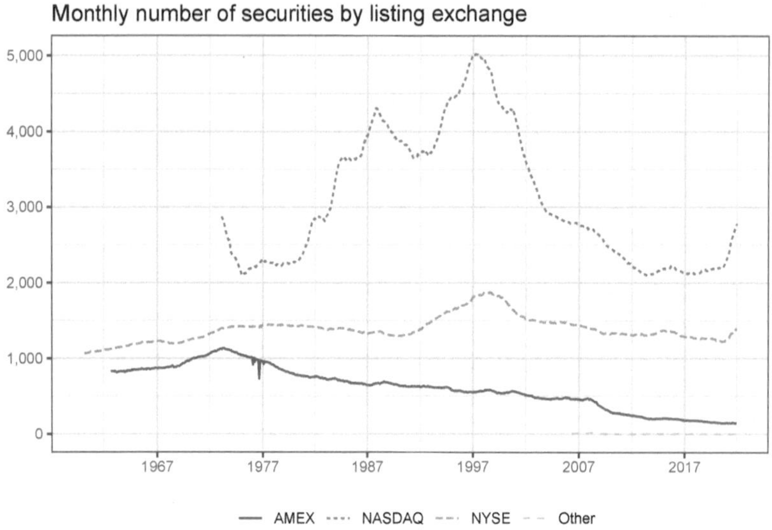

FIGURE 3.1
Number of stocks in the CRSP sample listed at each of the US exchanges.

```
) |>
collect() |>
mutate(month = ymd(month)) |>
ggplot(aes(
  x = month, y = mktcap / 1000,
  color = exchange, linetype = exchange
)) +
geom_line() +
labs(
  x = NULL, y = NULL, color = NULL, linetype = NULL,
  title = "Monthly market cap by listing exchange in billions of Dec 2021 USD"
) +
scale_x_date(date_breaks = "10 years", date_labels = "%Y") +
scale_y_continuous(labels = comma)
```

Of course, performing the computation in the database is not really meaningful because we can easily pull all the required data into our memory. The code chunk above is slower than performing the same steps on tables that are already in memory. However, we just want to illustrate that you can perform many things in the database before loading the data into your memory. Before we proceed, we load the monthly CPI data.

FIGURE 3.2
Market capitalization is measured in billion USD, adjusted for consumer price index changes such that the values on the horizontal axis reflect the buying power of billion USD in December 2021.

```
cpi_monthly <- tbl(tidy_finance, "cpi_monthly") |>
  collect()
```

Next, we look at the same descriptive statistics by industry. Figure 3.3 plots the number of stocks in the sample for each of the SIC industry classifiers. For most of the sample period, the largest share of stocks is in manufacturing, albeit the number peaked somewhere in the 90s. The number of firms associated with public administration seems to be the only category on the rise in recent years, even surpassing manufacturing at the end of our sample period.

```
crsp_monthly_industry <- crsp_monthly |>
  left_join(cpi_monthly, by = "month") |>
  group_by(month, industry) |>
  summarize(
    securities = n_distinct(permno),
    mktcap = sum(mktcap) / mean(cpi),
    .groups = "drop"
  )

crsp_monthly_industry |>
  ggplot(aes(
```

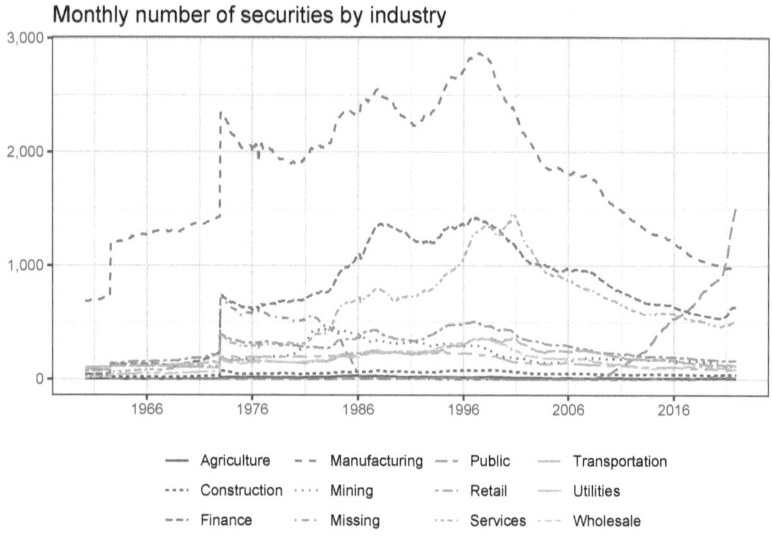

FIGURE 3.3
Number of stocks in the CRSP sample associated with different industries.

```
    x = month,
    y = securities,
    color = industry,
    linetype = industry
  )) +
  geom_line() +
  labs(
    x = NULL, y = NULL, color = NULL, linetype = NULL,
    title = "Monthly number of securities by industry"
  ) +
  scale_x_date(date_breaks = "10 years", date_labels = "%Y") +
  scale_y_continuous(labels = comma)
```

We also compute the market cap of all stocks belonging to the respective industries and show the evolution over time in Figure 3.4. All values are again in terms of billions of end of 2021 USD. At all points in time, manufacturing firms comprise of the largest portion of market capitalization. Toward the end of the sample, however, financial firms and services begin to make up a substantial portion of the market cap.

```
crsp_monthly_industry |>
  ggplot(aes(
    x = month,
    y = mktcap / 1000,
```

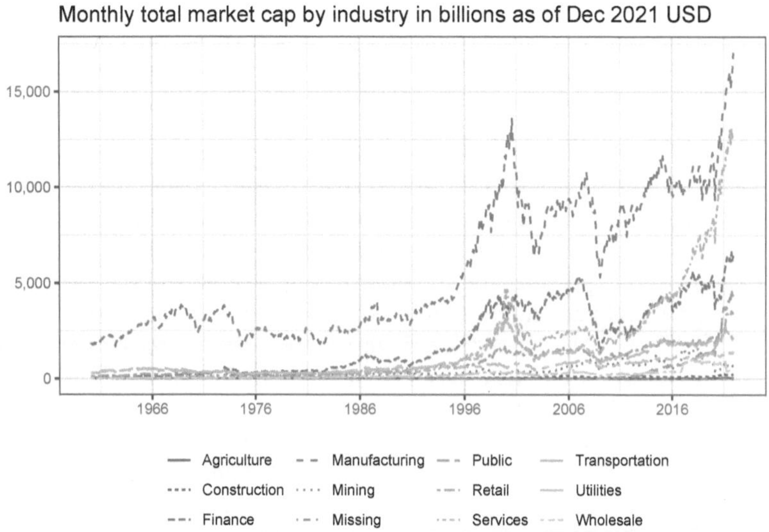

FIGURE 3.4
Market capitalization is measured in billion USD, adjusted for consumer price index changes such that the values on the y-axis reflect the buying power of billion USD in December 2021.

```
    color = industry,
    linetype = industry
  )) +
  geom_line() +
  labs(
    x = NULL, y = NULL, color = NULL, linetype = NULL,
    title = "Monthly total market cap by industry in billions as of Dec 2021 USD"
  ) +
  scale_x_date(date_breaks = "10 years", date_labels = "%Y") +
  scale_y_continuous(labels = comma)
```

3.4 Daily CRSP Data

Before we turn to accounting data, we provide a proposal for downloading daily CRSP data. While the monthly data from above typically fit into your memory and can be downloaded in a meaningful amount of time, this is usually not true for daily return data. The daily CRSP data file is substantially larger than monthly data and can exceed 20GB. This has two important implications: you cannot hold all the daily return data in your memory (hence it is not possible to copy the entire data set to your local database), and in our experience, the download usually crashes (or never stops) because it is too much data for the WRDS cloud to prepare and send to your R session.

There is a solution to this challenge. As with many *big data* problems, you can split up the big task into several smaller tasks that are easy to handle. That is, instead of downloading data about many stocks all at once, download the data in small batches for each stock consecutively. Such operations can be implemented in `for()`-loops, where we download, prepare, and store the data for a single stock in each iteration. This operation might nonetheless take a couple of hours, so you have to be patient either way (we often run such code overnight). To keep track of the progress, you can use `txtProgressBar()`. Eventually, we end up with more than 68 million rows of daily return data. Note that we only store the identifying information that we actually need, namely `permno`, `date`, and `month` alongside the excess returns. We thus ensure that our local database contains only the data we actually use and that we can load the full daily data into our memory later. Notice that we also use the function `dbWriteTable()` here with the option to append the new data to an existing table, when we process the second and all following batches.

```
dsf_db <- tbl(wrds, in_schema("crsp", "dsf"))

factors_ff_daily <- tbl(tidy_finance, "factors_ff_daily") |>
  collect()

permnos <- tbl(tidy_finance, "crsp_monthly") |>
  distinct(permno) |>
  pull()

progress <- txtProgressBar(
  min = 0,
  max = length(permnos),
  initial = 0,
  style = 3
)

for (j in 1:length(permnos)) {
  permno_sub <- permnos[j]
  crsp_daily_sub <- dsf_db |>
    filter(permno == permno_sub &
      date >= start_date & date <= end_date) |>
    select(permno, date, ret) |>
    collect() |>
    drop_na()

  if (nrow(crsp_daily_sub) > 0) {
```

```
    crsp_daily_sub <- crsp_daily_sub |>
      mutate(month = floor_date(date, "month")) |>
      left_join(factors_ff_daily |>
        select(date, rf), by = "date") |>
      mutate(
        ret_excess = ret - rf,
        ret_excess = pmax(ret_excess, -1)
      ) |>
      select(permno, date, month, ret_excess)

    dbWriteTable(tidy_finance,
        "crsp_daily",
        value = crsp_daily_sub,
        overwrite = ifelse(j == 1, TRUE, FALSE),
        append = ifelse(j != 1, TRUE, FALSE)
      )
  }
  setTxtProgressBar(progress, j)
}

close(progress)

crsp_daily_db <- tbl(tidy_finance, "crsp_daily")
```

3.5 Preparing Compustat Data

Firm accounting data are an important source of information that we use in portfolio analyses in subsequent chapters. The commonly used source for firm financial information is Compustat provided by S&P Global Market Intelligence,[5] which is a global data vendor that provides financial, statistical, and market information on active and inactive companies throughout the world. For US and Canadian companies, annual history is available back to 1950 and quarterly as well as monthly histories date back to 1962.

To access Compustat data, we can again tap WRDS, which hosts the `funda` table that contains annual firm-level information on North American companies.

```
funda_db <- tbl(wrds, in_schema("comp", "funda"))
```

[5]https://www.spglobal.com/marketintelligence/en/

We follow the typical filter conventions and pull only data that we actually need: (i) we get only records in industrial data format, (ii) in the standard format (i.e., consolidated information in standard presentation), and (iii) only data in the desired time window.

```
compustat <- funda_db |>
  filter(
    indfmt == "INDL" &
      datafmt == "STD" &
      consol == "C" &
      datadate >= start_date & datadate <= end_date
  ) |>
  select(
    gvkey, # Firm identifier
    datadate, # Date of the accounting data
    seq, # Stockholders' equity
    ceq, # Total common/ordinary equity
    at, # Total assets
    lt, # Total liabilities
    txditc, # Deferred taxes and investment tax credit
    txdb, # Deferred taxes
    itcb, # Investment tax credit
    pstkrv, # Preferred stock redemption value
    pstkl, # Preferred stock liquidating value
    pstk, # Preferred stock par value
    capx, # Capital investment
    oancf # Operating cash flow
  ) |>
  collect()
```

Next, we calculate the book value of preferred stock and equity inspired by the variable definition in Ken French's data library.[6] Note that we set negative or zero equity to missing which is a common practice when working with book-to-market ratios (see Fama and French, 1992, for details).

```
compustat <- compustat |>
  mutate(
    be = coalesce(seq, ceq + pstk, at - lt) +
      coalesce(txditc, txdb + itcb, 0) -
      coalesce(pstkrv, pstkl, pstk, 0),
    be = if_else(be <= 0, as.numeric(NA), be)
  )
```

[6]https://mba.tuck.dartmouth.edu/pages/faculty/ken.french/Data_Library/variable_definitions.html

We keep only the last available information for each firm-year group. Note that `datadate` defines the time the corresponding financial data refers to (e.g., annual report as of December 31, 2021). Therefore, `datadate` is not the date when data was made available to the public. Check out the exercises for more insights into the peculiarities of `datadate`.

```
compustat <- compustat |>
  mutate(year = year(datadate)) |>
  group_by(gvkey, year) |>
  filter(datadate == max(datadate)) |>
  ungroup()
```

With the last step, we are already done preparing the firm fundamentals. Thus, we can store them in our local database.

```
  dbWriteTable(tidy_finance,
    "compustat",
    value = compustat,
    overwrite = TRUE
  )
```

3.6 Merging CRSP with Compustat

Unfortunately, CRSP and Compustat use different keys to identify stocks and firms. CRSP uses `permno` for stocks, while Compustat uses `gvkey` to identify firms. Fortunately, a curated matching table on WRDS allows us to merge CRSP and Compustat, so we create a connection to the *CRSP-Compustat Merged* table (provided by CRSP).

```
ccmxpf_linktable_db <- tbl(
  wrds,
  in_schema("crsp", "ccmxpf_linktable")
)
```

The linking table contains links between CRSP and Compustat identifiers from various approaches. However, we need to make sure that we keep only relevant and correct links, again following the description outlined in Bali et al. (2016). Note also that currently active links have no end date, so we just enter the current date via `today()`.

```
ccmxpf_linktable <- ccmxpf_linktable_db |>
  filter(linktype %in% c("LU", "LC") &
    linkprim %in% c("P", "C") &
```

```
    usedflag == 1) |>
  select(permno = lpermno, gvkey, linkdt, linkenddt) |>
  collect() |>
  mutate(linkenddt = replace_na(linkenddt, today()))
```

We use these links to create a new table with a mapping between stock identifier, firm identifier, and month. We then add these links to the Compustat `gvkey` to our monthly stock data.

```
ccm_links <- crsp_monthly |>
  inner_join(ccmxpf_linktable, by = "permno") |>
  filter(!is.na(gvkey) & (date >= linkdt & date <= linkenddt)) |>
  select(permno, gvkey, date)

crsp_monthly <- crsp_monthly |>
  left_join(ccm_links, by = c("permno", "date"))
```

As the last step, we update the previously prepared monthly CRSP file with the linking information in our local database.

```
  dbWriteTable(tidy_finance,
    "crsp_monthly",
    value = crsp_monthly,
    overwrite = TRUE
  )
```

Before we close this chapter, let us look at an interesting descriptive statistic of our data. As the book value of equity plays a crucial role in many asset pricing applications, it is interesting to know for how many of our stocks this information is available. Hence, Figure 3.5 plots the share of securities with book equity values for each exchange. It turns out that the coverage is pretty bad for AMEX- and NYSE-listed stocks in the 60s but hovers around 80% for all periods thereafter. We can ignore the erratic coverage of securities that belong to the other category since there is only a handful of them anyway in our sample.

```
crsp_monthly |>
  group_by(permno, year = year(month)) |>
  filter(date == max(date)) |>
  ungroup() |>
  left_join(compustat, by = c("gvkey", "year")) |>
  group_by(exchange, year) |>
  summarize(
    share = n_distinct(permno[!is.na(be)]) / n_distinct(permno),
    .groups = "drop"
```

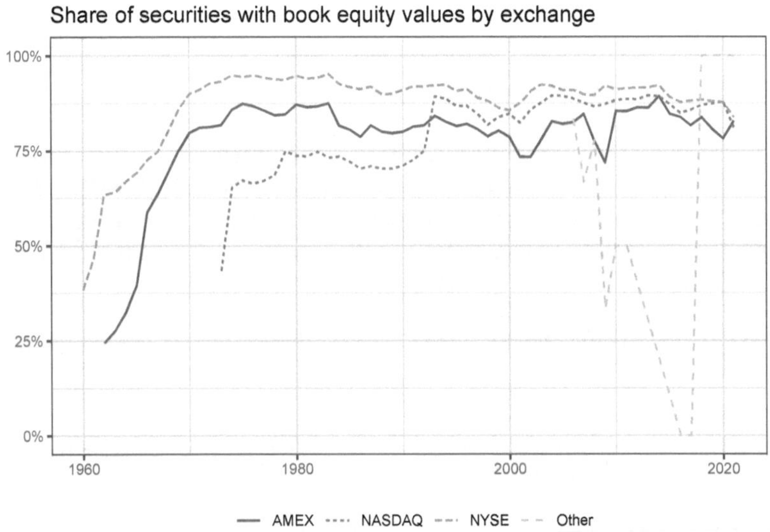

FIGURE 3.5
End-of-year share of securities with book equity values by listing exchange.

```
) |>
ggplot(aes(
  x = year,
  y = share,
  color = exchange,
  linetype = exchange
  )) +
geom_line() +
labs(
  x = NULL, y = NULL, color = NULL, linetype = NULL,
  title = "Share of securities with book equity values by exchange"
) +
scale_y_continuous(labels = percent) +
coord_cartesian(ylim = c(0, 1))
```

3.7 Some Tricks for PostgreSQL Databases

As we mentioned above, the WRDS database runs on PostgreSQL rather than SQLite. Finding the right tables for your data needs can be tricky in the WRDS PostgreSQL instance, as the tables are organized in schemas. If you wonder what the purpose of schemas is, check out this documetation.[7] For instance, if you want to find all tables that live in the `crsp` schema, you run

```
dbListObjects(wrds, Id(schema = "crsp"))
```

This operation returns a list of all tables that belong to the `crsp` family on WRDS, e.g., `<Id> schema = crsp, table = msenames`. Similarly, you can fetch a list of all tables that belong to the `comp` family via

```
dbListObjects(wrds, Id(schema = "comp"))
```

If you want to get all schemas, then run

```
dbListObjects(wrds)
```

3.8 Exercises

1. Check out the structure of the WRDS database by sending queries in the spirit of "Querying WRDS Data using R"[8] and verify the output with `dbListObjects()`. How many tables are associated with CRSP? Can you identify what is stored within *msp500*?
2. Compute `mkt_cap_lag` using `lag(mktcap)` rather than joins as above. Filter out all the rows where the lag-based market capitalization measure is different from the one we computed above. Why are they different?
3. In the main part, we look at the distribution of market capitalization across exchanges and industries. Now, plot the average market capitalization of firms for each exchange and industry. What do you find?
4. `datadate` refers to the date to which the fiscal year of a corresponding firm refers to. Count the number of observations in Compustat by *month* of this date variable. What do you find? What does the finding suggest about pooling observations with the same fiscal year?

[7] https://www.postgresql.org/docs/9.1/ddl-schemas.html

[8] https://wrds-www.wharton.upenn.edu/pages/support/programming-wrds/programming-r/querying-wrds-data-r/

5. Go back to the original Compustat data in `funda_db` and extract rows where the same firm has multiple rows for the same fiscal year. What is the reason for these observations?
6. Repeat the analysis of market capitalization for book equity, which we computed from the Compustat data. Then, use the matched sample to plot book equity against market capitalization. How are these two variables related?

4

TRACE and FISD

In this chapter, we dive into the US corporate bond market. Bond markets are far more diverse than stock markets, as most issuers have multiple bonds outstanding simultaneously with potentially very different indentures. This market segment is exciting due to its size (roughly 10 trillion USD outstanding), heterogeneity of issuers (as opposed to government bonds), market structure (mostly over-the-counter trades), and data availability. We introduce how to use bond characteristics from FISD and trade reports from TRACE and provide code to download and clean TRACE in R.

Many researchers study liquidity in the US corporate bond market (see, e.g., Bessembinder et al., 2006, Edwards et al. (2007), and O'Hara and Zhou (2021), among many others). We do not cover bond returns here, but you can compute them from TRACE data. Instead, we refer to studies on the topic such as Bessembinder et al. (2008), Bai et al. (2019), and Kelly et al. (2021) and a survey by Huang and Shi (2021). Moreover, WRDS includes bond returns computed from TRACE data at a monthly frequency.

The current chapter relies on this set of packages.

```
library(tidyverse)
library(lubridate)
library(dbplyr)
library(RSQLite)
library(RPostgres)
library(devtools)
```

Compared to previous chapters, we load the `devtools` package (Wickham et al., 2022d) to source code that we provided to the public via gist.[1]

[1]https://docs.github.com/en/get-started/writing-on-github/editing-and-sharing-content-with-gists/creating-gists

DOI: 10.1201/b23237-4

4.1 Bond Data from WRDS

Both bond databases we need are available on WRDS[2] to which we establish the `RPostgres` connection described in the previous chapter. Additionally, we connect to our local `SQLite`-database to store the data we download.

```
wrds <- dbConnect(
  Postgres(),
  host = "wrds-pgdata.wharton.upenn.edu",
  dbname = "wrds",
  port = 9737,
  sslmode = "require",
  user = Sys.getenv("user"),
  password = Sys.getenv("password")
)

tidy_finance <- dbConnect(
  SQLite(),
  "data/tidy_finance.sqlite",
  extended_types = TRUE
)
```

4.2 Mergent FISD

For research on US corporate bonds, the Mergent Fixed Income Securities Database (FISD) is the primary resource for bond characteristics. There is a detailed manual[3] on WRDS, so we only cover the necessary subjects here. FISD data comes in two main variants, namely, centered on issuers or issues. In either case, the most useful identifiers are CUSIPs.[4] 9-digit CUSIPs identify securities issued by issuers. The issuers can be identified from the first six digits of a security CUSIP, which is also called 6-digit CUSIP. Both stocks and bonds have CUSIPs. This connection would, in principle, allow matching them easily, but due to changing issuer details, this approach only yields small coverage.

[2] https://wrds-www.wharton.upenn.edu/

[3] https://wrds-www.wharton.upenn.edu/documents/1364/FixedIncome_Securities_Master_Database_User_Guide_v4.pdf

[4] https://www.cusip.com/index.html

We use the issue-centered version of FISD to identify the subset of US corporate bonds that meet the standard criteria (Bessembinder et al., 2006). The WRDS table `fisd_mergedissue` contains most of the information we need on a 9-digit CUSIP level. Due to the diversity of corporate bonds, details in the indenture vary significantly. We focus on common bonds that make up the majority of trading volume in this market without diverging too much in indentures.

The following chunk connects to the data and selects the bond sample to remove certain bond types that are less commonly (see, e.g., Dick-Nielsen et al., 2012; O'Hara and Zhou, 2021, among many others).

```
mergent <- tbl(
  wrds,
  in_schema("fisd", "fisd_mergedissue")
) |>
  filter(
    security_level == "SEN", # senior bonds
    slob == "N", # secured lease obligation
    is.na(security_pledge), # unsecured bonds
    asset_backed == "N", # not asset backed
    defeased == "N", # not defeased
    bond_type %in% c(
      "CDEB", # US Corporate Debentures
      "CMTN", # US Corporate MTN (Medium Term Note)
      "CMTZ", # US Corporate MTN Zero
      "CZ", # US Corporate Zero,
      "USBN" # US Corporate Bank Note
    ),
    pay_in_kind != "Y", # not payable in kind
    yankee == "N", # no foreign issuer
    canadian == "N", # not Canadian
    foreign_currency == "N", # USD
    coupon_type %in% c(
      "F", # fixed coupon
      "Z" # zero coupon
    ),
    is.na(fix_frequency),
    coupon_change_indicator == "N",
    interest_frequency %in% c(
      "0", # per year
      "1",
      "2",
      "4",
      "12"
    ),
    rule_144a == "N", # publicly traded
    private_placement == "N",
```

```
    defaulted == "N", # not defaulted
    is.na(filing_date),
    is.na(settlement),
    convertible == "N", # not convertible
    is.na(exchange),
    putable == "N", # not putable
    unit_deal == "N", # not issued with another security
    exchangeable == "N", # not exchangeable
    perpetual == "N", # not perpetual
    preferred_security == "N" # not preferred
  ) |>
  select(
    complete_cusip, maturity,
    offering_amt, offering_date,
    dated_date,
    interest_frequency, coupon,
    last_interest_date,
    issue_id, issuer_id
  ) |>
  collect()
```

We also pull issuer information from `fisd_mergedissuer` regarding the industry and country of the firm that issued a particular bond. Then, we filter to include only US-domiciled firms' bonds. We match the data by `issuer_id`.

```
mergent_issuer <- tbl(wrds, in_schema("fisd", "fisd_mergedissuer")) |>
  select(issuer_id, sic_code, country_domicile) |>
  collect()

mergent <- mergent |>
  inner_join(mergent_issuer, by = "issuer_id") |>
  filter(country_domicile == "USA") |>
  select(-country_domicile)
```

Finally, we save the bond characteristics to our local database. This selection of bonds also constitutes the sample for which we will collect trade reports from TRACE below.

```
  dbWriteTable(
    conn = tidy_finance,
    name = "mergent",
    value = mergent,
    overwrite = TRUE
  )
```

The FISD database also contains other data. The issue-based file contains information on covenants, i.e., restrictions included in bond indentures to limit specific actions by firms (e.g., Handler et al., 2021). Moreover, FISD also provides information on bond ratings. We do not need either here.

4.3 TRACE

The Financial Industry Regulatory Authority (FINRA) provides the Trade Reporting and Compliance Engine (TRACE). In TRACE, dealers that trade corporate bonds must report such trades individually. Hence, we observe trade messages in TRACE that contain information on the bond traded, the trade time, price, and volume. TRACE comes in two variants; standard and enhanced TRACE. We show how to download and clean enhanced TRACE as it contains uncapped volume, a crucial quantity missing in the standard distribution. Moreover, enhanced TRACE also provides information on the respective parties' roles and the direction of the trade report. These items become essential in cleaning the messages.

Why do we repeatedly talk about cleaning TRACE? Trade messages are submitted within a short time window after a trade is executed (less than 15 minutes). These messages can contain errors, and the reporters subsequently correct them or they cancel a trade altogether. The cleaning needs are described by Dick-Nielsen (2009) in detail, and Dick-Nielsen (2014) shows how to clean the enhanced TRACE data using SAS. We do not go into the cleaning steps here, since the code is lengthy and serves no educational purpose. However, downloading and cleaning enhanced TRACE data is straightforward with our setup.

We store code for cleaning enhanced TRACE with R on the following Github gist.[5] as a function. The appendix also contains the code for reference. We only need to source the code from the gist, which we can do with `source_gist()`. Alternatively, you can also go to the gist, download it, and `source()` the respective R-file. The `clean_enhanced_trace()` function takes a vector of CUSIPs, a connection to WRDS explained in Chapter 3, and a start and end date, respectively.

```
source_gist("3a05b3ab281563b2e94858451c2eb3a4")
```

The TRACE database is considerably large. Therefore, we only download subsets of data at once. Specifying too many CUSIPs over a long time horizon will result in very long download times and a potential failure due to the size of the request to WRDS. The size limit depends on many parameters, and we cannot give you a guideline here. If we were working with the complete TRACE data for all CUSIPs above, splitting the

[5]https://gist.github.com/patrick-weiss/3a05b3ab281563b2e94858451c2eb3a4

data into 100 parts takes roughly two hours using our setup. For the applications in this book, we need data around the Paris Agreement in December 2015 and download the data in ten sets, which we define below.

```
mergent_cusips <- mergent |>
  pull(complete_cusip)

mergent_parts <- split(
  mergent_cusips,
  rep(1:10,
    length.out = length(mergent_cusips))
)
```

Finally, we run a loop in the same style as in Chapter 3 where we download daily returns from CRSP. For each of the CUSIP sets defined above, we call the cleaning function and save the resulting output. We add new data to the existing table for batch two and all following batches.

```
for (j in 1:length(mergent_parts)) {
  trace_enhanced <- clean_enhanced_trace(
    cusips = mergent_parts[[j]],
    connection = wrds,
    start_date = ymd("2014-01-01"),
    end_date = ymd("2016-11-30")
  )

  dbWriteTable(
    conn = tidy_finance,
    name = "trace_enhanced",
    value = trace_enhanced,
    overwrite = ifelse(j == 1, TRUE, FALSE),
    append = ifelse(j != 1, TRUE, FALSE)
  )
}
```

4.4 Insights into Corporate Bonds

While many news outlets readily provide information on stocks and the underlying firms, corporate bonds are not covered frequently. Additionally, the TRACE database contains trade-level information, potentially new to students. Therefore, we provide you with some insights by showing some summary statistics.

We start by looking into the number of bonds outstanding over time and compare it to the number of bonds traded in our sample. First, we compute the number of bonds outstanding for each quarter around the Paris Agreement from 2014 to 2016.

```
bonds_outstanding <- expand_grid("date" = seq(ymd("2014-01-01"),
                                              ymd("2016-11-30"),
                                              by = "quarter"),
                                 "complete_cusip" = mergent$complete_cusip) |>
  left_join(mergent |> select(complete_cusip,
                              offering_date,
                              maturity),
            by = "complete_cusip") |>
  mutate(offering_date = floor_date(offering_date),
         maturity = floor_date(maturity)) |>
  filter(date >= offering_date & date <= maturity) |>
  count(date) |>
  mutate(type = "Outstanding")
```

Next, we look at the bonds traded each quarter in the same period. Notice that we load the complete trace table from our database, as we only have a single part of it in the environment from the download loop from above.

```
trace_enhanced <- tbl(tidy_finance, "trace_enhanced") |>
  collect()

bonds_traded <- trace_enhanced |>
  mutate(date = floor_date(trd_exctn_dt, "quarters")) |>
  group_by(date) |>
  summarize(n = length(unique(cusip_id)),
            type = "Traded",
            .groups = "drop")
```

Finally, we plot the two time series in Figure 4.1.

```
bonds_outstanding |>
  bind_rows(bonds_traded) |>
  ggplot(aes(
    x = date,
    y = n,
    color = type,
    linetype = type
  )) +
  geom_line() +
  labs(
    x = NULL, y = NULL, color = NULL, linetype = NULL,
```

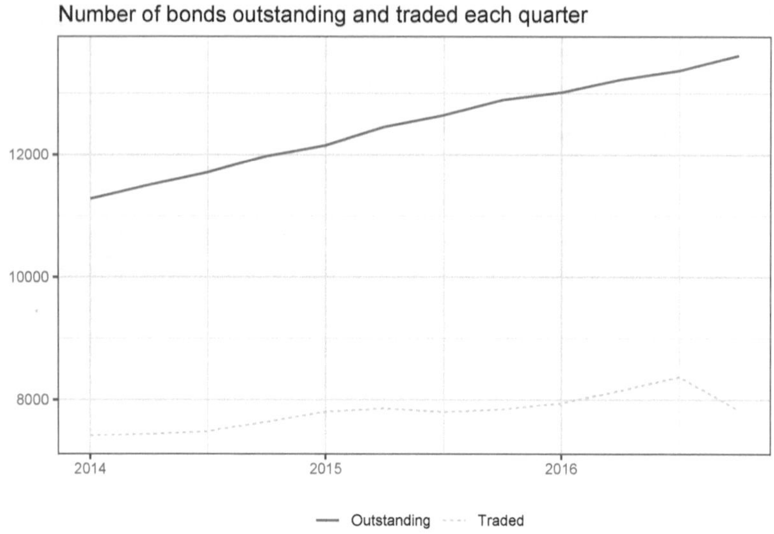

FIGURE 4.1
The number of corporate bonds outstanding each quarter as reported by Mergent FISD and the number of traded bonds from enhanced TRACE between 2014 and end of 2016.

```
    title = "Number of bonds outstanding and traded each quarter"
  )
```

We see that the number of bonds outstanding increases steadily between 2014 and 2016. During our sample period of trade data, we see that the fraction of bonds trading each quarter is roughly 60%. The relatively small number of traded bonds means that many bonds do not trade through an entire quarter. This lack of trading activity illustrates the generally low level of liquidity in the corporate bond market, where it can be hard to trade specific bonds. Does this lack of liquidity mean that corporate bond markets are irrelevant in terms of their size? With over 7,500 traded bonds each quarter, it is hard to say that the market is small. However, let us also investigate the characteristics of issued corporate bonds. In particular, we consider maturity (in years), coupon, and offering amount (in million USD).

```
mergent |>
  mutate(maturity = as.numeric(maturity - offering_date) / 365,
         offering_amt = offering_amt / 10^3) |>
  pivot_longer(cols = c(maturity, coupon, offering_amt),
               names_to = "measure") |>
  drop_na() |>
  group_by(measure) |>
```

```
  summarize(
    mean = mean(value),
    sd = sd(value),
    min = min(value),
    q05 = quantile(value, 0.05),
    q50 = quantile(value, 0.50),
    q95 = quantile(value, 0.95),
    max = max(value)
  )
```

```
# A tibble: 3 x 8
  measure         mean     sd   min    q05    q50     q95    max
  <chr>          <dbl>  <dbl> <dbl>  <dbl>  <dbl>   <dbl>  <dbl>
1 coupon          6.21   2.43 0       2.25   6.38    9.86     39
2 maturity       10.1    9.13 0.252 1.26    8.02   30.0     100.
3 offering_amt 357.    545.   0.001 0.875 200     1250    15000
```

We see that the average bond in our sample period has an offering amount of over 357 million USD with a median of 200 million USD, which both cannot be considered small. The average bond has a maturity of 10 years and pays around 6% in coupons.

Finally, let us compute some summary statistics for the trades in this market. To this end, we show a summary based on aggregate information daily. In particular, we consider the trade size (in million USD) and the number of trades.

```
trace_enhanced |>
  group_by(trd_exctn_dt) |>
  summarize(trade_size = sum(entrd_vol_qt * rptd_pr / 100) / 10^6,
            trade_number = n(),
            .groups = "drop") |>
  pivot_longer(cols = c(trade_size, trade_number),
               names_to = "measure") |>
  group_by(measure) |>
  summarize(
    mean = mean(value),
    sd = sd(value),
    min = min(value),
    q05 = quantile(value, 0.05),
    q50 = quantile(value, 0.50),
    q95 = quantile(value, 0.95),
    max = max(value)
  )
```

```
# A tibble: 2 x 8
  measure         mean     sd   min    q05    q50     q95    max
```

```
  <chr>            <dbl> <dbl> <dbl>  <dbl>  <dbl>  <dbl>  <dbl>
1 trade_number 25921. 5460. 438    17851. 26025  34458. 40889
2 trade_size   12968. 3574.  17.2  6138. 13408. 17851. 20905.
```

On average, nearly 26 billion USD of corporate bonds are traded daily in nearly 13,000 transactions. We can hence conclude that the corporate bond market is indeed significant in terms of trading volume and activity.

4.5 Exercises

1. Summarize the amount outstanding of all bonds over time and describe the resulting graph.
2. Compute the number of days each bond is traded (accounting for the bonds' maturities and issuances). Start by looking at the number of bonds traded each day in a graph similar to the one above. How many bonds trade on more than 75% of trading days?
3. WRDS provides more information from Mergent FISD. In particular, they also provide rating information in `fisd_ratings`. Download the ratings for the bond sample. Then, plot the distribution of ratings in a histogram.
4. Download the TRACE data until the end of September 2021. Hint: you want to download the data completely new, rather than adding the new information. Can you find a reason why?

5

Other Data Providers

In the previous chapters, we introduced many ways to get financial data that researchers regularly use. We showed how to load data into R from Yahoo!Finance and commonly used file types, such as comma-separated or Excel files. Then, we introduced remotely connecting to WRDS and downloading data from there. However, this is only a subset of the vast amounts of data available these days.

In this short chapter, we aim to provide an overview of common alternative data providers for which direct access via R packages exists. Such a list requires constant adjustments because both data providers and access methods change. However, we want to emphasize two main insights: First, the number of R packages that provide access to (financial) data is large. Too large actually to survey here exhaustively. Instead, we can only cover the tip of the iceberg. Second, R provides the functionalities to access basically any form of files or data available online. Thus, even if a desired data source does not come with a well-established R package, chances are high that data can be retrieved by establishing your own API connection or by scrapping the content.

In our non-exhaustive list below, we restrict ourselves to listing data sources accessed through easy-to-use R packages. For further inspiration on potential data sources, we recommend reading the R task view empirical finance.[1] Further inspiration (on more general social sciences) can be found here.[2]

Source	Description	R packages
	Macroeconomic Variables	
FED	The Federal Reserve Bank of St. Louis provides more than 818,000 US and international time series from 109 sources via the API FRED. The data is freely available and can be browsed online on the FRED homepage.[3]	`fredr` (Boysel and Vaughan, 2021) and `alfred` (Kleen, 2021)

[1]https://cran.r-project.org/web/views/Finance.html
[2]https://cengel.github.io/gearup2016/SULdataAccess.html

DOI: 10.1201/b23237-5

Source	Description	R packages
ECB	The European Central Bank's Statistical Data Warehouse[4] provides data on Euro area monetary policy, financial stability, and other topics relevant to the activities of the ECB and the European System of Central Banks (ESCB).	`ecb` (Persson, 2021)
	Financial data	
Bloomberg[5]	Bloomberg's Fundamental coverage includes current and normalized historical data for the balance sheet, income statement, cash flows statement, and financial ratios. Additionally, it provides industry-specific data for communications, consumer, energy, health care, and many more. In order to retrieve Bloomberg data, a paid subscription is needed.	`Rblpapi` (Armstrong et al., 2022)
Refinitiv Eikon[6]	Eikon provides access to real-time market data, news, fundamental data, analytics, trading, and messaging tools. Refinitiv's Eikon is a paid service. Apart from the CRAN version, there is also `https://github.com/philaris/eikonapir`.	`DatastreamDSWS2R` (Cara, 2021) and `eikonapir`
Nasdaq Data Link (Quandl)[7]	Quandl is a publisher of alternative data. Quandl publishes free data, scraped from many different sources from the web. However, some of the data requires specific subscriptions on the Quandl platform.	`Quandl` (McTaggart et al., 2021)
Global factor data	The data repository of Jensen et al. (2022b). They provide return data for characteristic-managed portfolios from around the world. The database includes factors for 153 characteristics in 13 themes, using data from 93 countries. Download the data here.[8]	
Open Source Asset Pricing	The data repository of Chen and Zimmermann (2022). They provide return data for over 200 trading strategies with different time periods and specifications. The authors also provide signals and explanations of the factor construction. Download the data here.[9]	
Simfin[10]	Simfin make fundamental financial data freely available to private investors, researchers, and students. The data provider applies automating data collection processes to collect a large set of publicly available information from firms' financial statements.	`simfinapi` (Gomolka, 2021)
	High-frequency data	

Source	Description	R packages
IEX	The IEX Group operates the Investors Exchange (IEX), a stock exchange for US equities. IEX offers US reference and market data including end-of-day and *intraday pricing data*. IEX offers an API which is freely available.	`Riex` (Ibrahim, 2021)
TAQ	TAQ data provides subscribed users access to all trades and quotes for all issues traded on NYSE, Nasdaq, and the regional exchanges. TAQ data[11] can be accessed from WRDS via Postgres. The `highfrequency` package delivers useful workflows to clean TAQ data.	`highfrequency` (Boudt et al., 2022)
	Other (free) data	
Crypto data	The data provider coinmarketcap[12] retrieves cryptocurrency information and historical prices as well as information on the exchanges they are listed on.	`crypto2` (Stoeckl, 2022)
Twitter	Twitter provides (limited) access for academic research to extract and analyze Tweets.	`rtweet` (Kearney, 2019)
SEC company fillings	The EDGAR[13] database provides free public access to corporate information, allowing you to research a public company's financial information and operations by reviewing the filings the company makes with the SEC. You can also research information provided by mutual funds (including money market funds), exchange-traded funds (ETFs), and variable annuities.	`edgarWebR` (Waldstein, 2021)
Google trends	Google offers public access to global search volumes through its search engine through the Google Trends portal.[14]	`globaltrends` (Puhr and Müllner, 2021) and `gtrends` (Massicotte and Eddelbuettel, 2022)

[3] https://fred.stlouisfed.org/
[4] https://sdw.ecb.europa.eu/
[5] https://www.bloomberg.com/
[6] https://www.refinitiv.com/en/financial-data
[7] data.nasdaq.com/publishers/qdl
[8] https://jkpfactors.com/
[9] https://www.openassetpricing.com/data
[10] https://simfin.com/
[11] https://www.nyse.com/market-data/historical
[12] coinmarketcap.com
[13] https://www.sec.gov/edgar/about
[14] https://trends.google.com/trends/?geo=DK

5.1 Exercises

1. Select one of the data sources in the table above and retrieve some data: Browse the homepage of the data provider or the package documentation to find inspiration on which type of data is available to you and how to download the data into your R session.
2. Generate summary statistics of the data you retrieved and provide some useful visualization. The possibilities are endless: Maybe there is some interesting economic event you want to analyze, such as stock market responses to Twitter activity.
3. Simfin[15] provides excellent data coverage. Use their API to find out if the information Simfin provides overlaps with the CRSP/Compustat dataset in the `tidy_finance.sqlite` database introduced in Chapters 2–4.

[15]https://simfin.com/

Part III

Asset Pricing

6

Beta Estimation

In this chapter, we introduce an important concept in financial economics: the exposure of an individual stock to changes in the market portfolio. According to the Capital Asset Pricing Model (CAPM) of Sharpe (1964), Lintner (1965), and Mossin (1966), cross-sectional variation in expected asset returns should be a function of the covariance between the excess return of the asset and the excess return on the market portfolio. The regression coefficient of excess market returns on excess stock returns is usually called the market beta. We show an estimation procedure for the market betas. We do not go into details about the foundations of market beta but simply refer to any treatment of the CAPM[1] for further information. Instead, we provide details about all the functions that we use to compute the results. In particular, we leverage useful computational concepts: rolling-window estimation and parallelization.

We use the following packages throughout this chapter:

```
library(tidyverse)
library(RSQLite)
library(scales)
library(slider)
library(furrr)
```

Compared to previous chapters, we introduce `slider` (Vaughan, 2021) for sliding window functions, and `furrr` (Vaughan and Dancho, 2022) to apply mapping functions in parallel.

6.1 Estimating Beta using Monthly Returns

The estimation procedure is based on a rolling-window estimation, where we may use either monthly or daily returns and different window lengths. First, let us start with loading the monthly CRSP data from our `SQLite`-database introduced in the previous Chapters 2–4.

[1]https://en.wikipedia.org/wiki/Capital_asset_pricing_model

DOI: 10.1201/b23237-6

```
tidy_finance <- dbConnect(
  SQLite(),
  "data/tidy_finance.sqlite",
  extended_types = TRUE
)

crsp_monthly <- tbl(tidy_finance, "crsp_monthly") |>
  collect()

factors_ff_monthly <- tbl(tidy_finance, "factors_ff_monthly") |>
  collect()

crsp_monthly <- crsp_monthly |>
  left_join(factors_ff_monthly, by = "month") |>
  select(permno, month, industry, ret_excess, mkt_excess)
```

To estimate the CAPM regression coeffients

$$r_{i,t} - r_{f,t} = \alpha_i + \beta_i(r_{m,t} - r_{f,t}) + \varepsilon_{i,t}$$

we regress stock excess returns `ret_excess` on excess returns of the market portfolio `mkt_excess`. R provides a simple solution to estimate (linear) models with the function `lm()`. `lm()` requires a formula as input that is specified in a compact symbolic form. An expression of the form `y ~ model` is interpreted as a specification that the response `y` is modeled by a linear predictor specified symbolically by `model`. Such a model consists of a series of terms separated by `+` operators. In addition to standard linear models, `lm()` provides a lot of flexibility. You should check out the documentation for more information. To start, we restrict the data only to the time series of observations in CRSP that correspond to Apple's stock (i.e., to `permno` 14593 for Apple) and compute $\hat{\alpha}_i$ as well as $\hat{\beta}_i$.

```
fit <- lm(ret_excess ~ mkt_excess,
  data = crsp_monthly |>
    filter(permno == "14593")
)

summary(fit)
```

```
Call:
lm(formula = ret_excess ~ mkt_excess, data = filter(crsp_monthly,
    permno == "14593"))

Residuals:
    Min      1Q  Median      3Q     Max 
-0.5169 -0.0598  0.0001  0.0636  0.3944 
```

```
Coefficients:
            Estimate Std. Error t value Pr(>|t|)
(Intercept)  0.01034    0.00521    1.99    0.048 *
mkt_excess   1.39419    0.11576   12.04   <2e-16 ***
---
Signif. codes:  0 '***' 0.001 '**' 0.01 '*' 0.05 '.' 0.1 ' ' 1

Residual standard error: 0.114 on 490 degrees of freedom
Multiple R-squared:  0.228, Adjusted R-squared:  0.227
F-statistic:  145 on 1 and 490 DF,  p-value: <2e-16
```

`lm()` returns an object of class `lm` which contains all information we usually care about with linear models. `summary()` returns an overview of the estimated parameters. `coefficients(fit)` would return only the estimated coefficients. The output above indicates that Apple moves excessively with the market as the estimated $\hat{\beta}_i$ is above one ($\hat{\beta}_i \approx 1.4$).

6.2 Rolling-Window Estimation

After we estimated the regression coefficients on an example, we scale the estimation of β_i to a whole different level and perform rolling-window estimations for the entire CRSP sample. The following function implements the CAPM regression for a data frame (or a part thereof) containing at least `min_obs` observations to avoid huge fluctuations if the time series is too short. If the condition is violated, that is, the time series is too short, the function returns a missing value.

```
estimate_capm <- function(data, min_obs = 1) {
  if (nrow(data) < min_obs) {
    beta <- as.numeric(NA)
  } else {
    fit <- lm(ret_excess ~ mkt_excess, data = data)
    beta <- as.numeric(coefficients(fit)[2])
  }
  return(beta)
}
```

Next, we define a function that does the rolling estimation. The `slide_period` function is able to handle months in its window input in a straightforward manner. We thus avoid using any time-series package (e.g., `zoo`) and converting the data to fit the package functions, but rather stay in the world of the `tidyverse`.

The following function takes input data and slides across the `month` vector, considering only a total of `months` months. The function essentially performs three steps: (i) arrange all rows, (ii) compute betas by sliding across months, and (iii) return a tibble with months and corresponding beta estimates (again particularly useful in the case of daily data). As we demonstrate further below, we can also apply the same function to daily returns data.

```
roll_capm_estimation <- function(data, months, min_obs) {
  data <- data |>
    arrange(month)

  betas <- slide_period_vec(
    .x = data,
    .i = data$month,
    .period = "month",
    .f = ~ estimate_capm(., min_obs),
    .before = months - 1,
    .complete = FALSE
  )

  return(tibble(
    month = unique(data$month),
    beta = betas
  ))
}
```

Before we attack the whole CRSP sample, let us focus on a couple of examples for well-known firms.

```
examples <- tribble(
  ~permno, ~company,
  14593, "Apple",
  10107, "Microsoft",
  93436, "Tesla",
  17778, "Berkshire Hathaway"
)
```

If we want to estimate rolling betas for Apple, we can use `mutate()`. We take a total of 5 years of data and require at least 48 months with return data to compute our betas. Check out the exercises if you want ot compute beta for different time periods.

```
beta_example <- crsp_monthly |>
  filter(permno == examples$permno[1]) |>
  mutate(roll_capm_estimation(cur_data(), months = 60, min_obs = 48)) |>
```

```
  drop_na()
beta_example
```

```
# A tibble: 445 x 6
  permno month      industry      ret_excess mkt_excess  beta
   <dbl> <date>     <chr>              <dbl>      <dbl> <dbl>
1  14593 1984-12-01 Manufacturing     0.170      0.0184  2.05
2  14593 1985-01-01 Manufacturing    -0.0108     0.0799  1.90
3  14593 1985-02-01 Manufacturing    -0.152      0.0122  1.88
4  14593 1985-03-01 Manufacturing    -0.112     -0.0084  1.89
5  14593 1985-04-01 Manufacturing    -0.0467    -0.0096  1.90
# ... with 440 more rows
```

It is actually quite simple to perform the rolling-window estimation for an arbitrary number of stocks, which we visualize in the following code chunk and the resulting Figure 6.1.

```
beta_examples <- crsp_monthly |>
  inner_join(examples, by = "permno") |>
  group_by(permno) |>
  mutate(roll_capm_estimation(cur_data(), months = 60, min_obs = 48)) |>
  ungroup() |>
  select(permno, company, month, beta) |>
  drop_na()

beta_examples |>
  ggplot(aes(
    x = month,
    y = beta,
    color = company,
    linetype = company)) +
  geom_line() +
  labs(
    x = NULL, y = NULL, color = NULL, linetype = NULL,
    title = "Monthly beta estimates for example stocks using 5 years of data"
  )
```

6.3 Parallelized Rolling-Window Estimation

Even though we could now just apply the function using `group_by()` on the whole CRSP sample, we advise against doing it as it is computationally quite expensive.

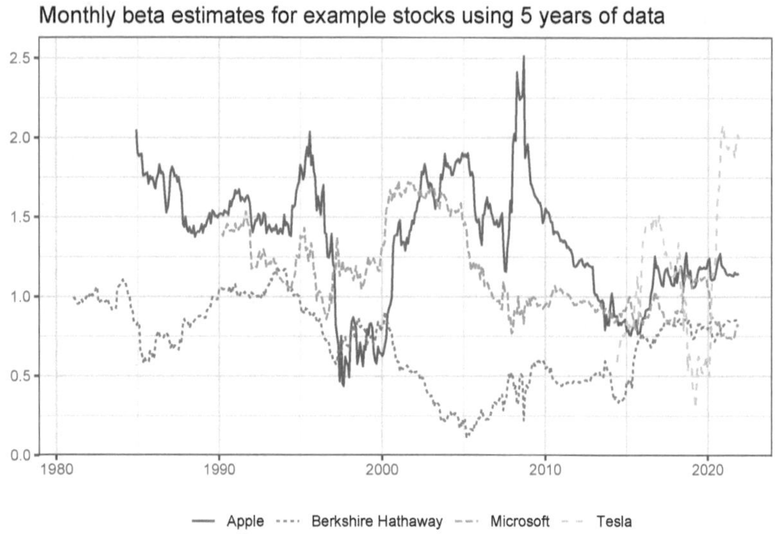

FIGURE 6.1
The CAPM betas are estimated with monthly data and a rolling window of length 5 years based on adjusted excess returns from CRSP. We use market excess returns from Kenneth French data library.

Remember that we have to perform rolling-window estimations across all stocks and time periods. However, this estimation problem is an ideal scenario to employ the power of parallelization. Parallelization means that we split the tasks which perform rolling-window estimations across different workers (or cores on your local machine).

First, we `nest()` the data by `permno`. Nested data means we now have a list of `permno` with corresponding time series data and an `industry` label. We get one row of output for each unique combination of non-nested variables which are `permno` and `industry`.

```
crsp_monthly_nested <- crsp_monthly |>
  nest(data = c(month, ret_excess, mkt_excess))
crsp_monthly_nested
```

```
# A tibble: 30,071 x 3
  permno industry      data
   <dbl> <chr>         <list>
1  10000 Manufacturing <tibble [16 x 3]>
2  10001 Utilities     <tibble [378 x 3]>
3  10002 Finance       <tibble [324 x 3]>
4  10003 Finance       <tibble [118 x 3]>
5  10005 Mining        <tibble [65 x 3]>
# ... with 30,066 more rows
```

Alternatively, we could have created the same nested data by *excluding* the variables that we *do not* want to nest, as in the following code chunk. However, for many applications it is desirable to explicitly state the variables that are nested into the `data` list-column, so that the reader can track what ends up in there.

```
crsp_monthly_nested <- crsp_monthly |>
  nest(data = -c(permno, industry))
```

Next, we want to apply the `roll_capm_estimation()` function to each stock. This situation is an ideal use case for `map()`, which takes a list or vector as input and returns an object of the same length as the input. In our case, `map()` returns a single data frame with a time series of beta estimates for each stock. Therefore, we use `unnest()` to transform the list of outputs to a tidy data frame.

```
crsp_monthly_nested |>
  inner_join(examples, by = "permno") |>
  mutate(beta = map(
    data,
    ~ roll_capm_estimation(., months = 60, min_obs = 48)
  )) |>
  unnest(beta) |>
  select(permno, month, beta_monthly = beta) |>
  drop_na()
```

```
# A tibble: 1,410 x 3
  permno month      beta_monthly
   <dbl> <date>            <dbl>
1  10107 1990-03-01         1.39
2  10107 1990-04-01         1.38
3  10107 1990-05-01         1.43
4  10107 1990-06-01         1.43
5  10107 1990-07-01         1.45
# ... with 1,405 more rows
```

However, instead, we want to perform the estimations of rolling betas for different stocks in parallel. If you have a Windows machine, it makes most sense to define `multisession`, which means that separate R processes are running in the background on the same machine to perform the individual jobs. If you check out the documentation of `plan()`, you can also see other ways to resolve the parallelization in different environments.

```
plan(multisession, workers = availableCores())
```

Using eight cores, the estimation for our sample of around 25k stocks takes around 20 minutes. Of course, you can speed up things considerably by having more cores available to share the workload or by having more powerful cores. Notice the difference in the code below? All you need to do is to replace `map()` with `future_map()`.

```
beta_monthly <- crsp_monthly_nested |>
  mutate(beta = future_map(
    data, ~ roll_capm_estimation(., months = 60, min_obs = 48)
  )) |>
  unnest(c(beta)) |>
  select(permno, month, beta_monthly = beta) |>
  drop_na()
```

6.4 Estimating Beta using Daily Returns

Before we provide some descriptive statistics of our beta estimates, we implement the estimation for the daily CRSP sample as well. Depending on the application, you might either use longer horizon beta estimates based on monthly data or shorter horizon estimates based on daily returns.

First, we load daily CRSP data. Note that the sample is large compared to the monthly data, so make sure to have enough memory available.

```
crsp_daily <- tbl(tidy_finance, "crsp_daily") |>
  collect()
```

We also need the daily Fama-French market excess returns.

```
factors_ff_daily <- tbl(tidy_finance, "factors_ff_daily") |>
  collect()
```

We make sure to keep only relevant data to save memory space. However, note that your machine might not have enough memory to read the whole daily CRSP sample. In this case, we refer you to the exercises and try working with loops as in Chapter 3.

```
crsp_daily <- crsp_daily |>
  inner_join(factors_ff_daily, by = "date") |>
  select(permno, month, ret_excess, mkt_excess)
```

Just like above, we nest the data by `permno` for parallelization.

```
crsp_daily_nested <- crsp_daily |>
  nest(data = c(month, ret_excess, mkt_excess))
```

This is what the estimation looks like for a couple of examples using `map()`. For the daily data, we use the same function as above but only take 3 months of data and require at least 50 daily return observations in these months. These restrictions help us to retrieve somewhat smooth coefficient estimates.

```
crsp_daily_nested |>
  inner_join(examples, by = "permno") |>
  mutate(beta_daily = map(
    data,
    ~ roll_capm_estimation(., months = 3, min_obs = 50)
  )) |>
  unnest(c(beta_daily)) |>
  select(permno, month, beta_daily = beta) |>
  drop_na()
```

```
# A tibble: 1,591 x 3
  permno month      beta_daily
   <dbl> <date>          <dbl>
1  10107 1986-05-01      0.898
2  10107 1986-06-01      0.906
3  10107 1986-07-01      0.822
4  10107 1986-08-01      0.900
5  10107 1986-09-01      1.01
# ... with 1,586 more rows
```

For the sake of completeness, we tell our session again to use multiple workers for parallelization.

```
plan(multisession, workers = availableCores())
```

The code chunk for beta estimation using daily returns now looks very similar to the one for monthly data. The whole estimation takes around 30 minutes using eight cores and 16gb memory.

```
beta_daily <- crsp_daily_nested |>
  mutate(beta_daily = future_map(
    data, ~ roll_capm_estimation(., months = 3, min_obs = 50)
  )) |>
  unnest(c(beta_daily)) |>
  select(permno, month, beta_daily = beta) |>
  drop_na()
```

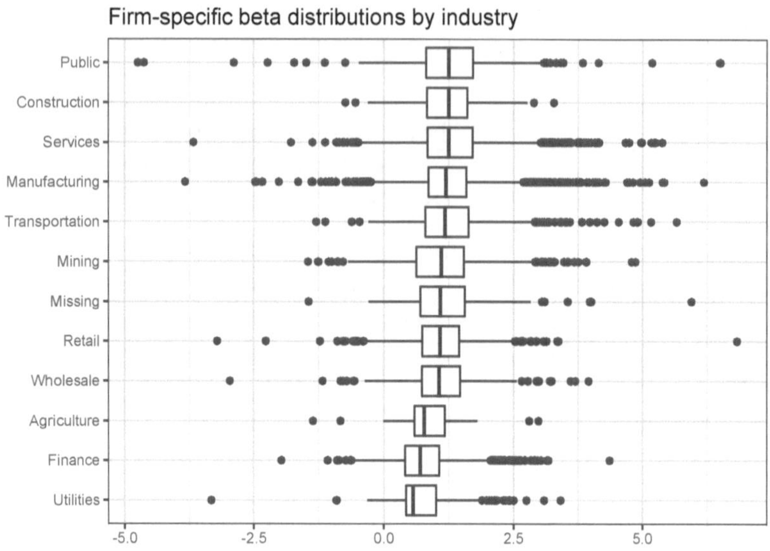

FIGURE 6.2
The box plots show the average firm-specific beta estimates by industry.

6.5 Comparing Beta Estimates

What is a typical value for stock betas? To get some feeling, we illustrate the dispersion of the estimated $\hat{\beta}_i$ across different industries and across time below. Figure 6.2 shows that typical business models across industries imply different exposure to the general market economy. However, there are barely any firms that exhibit a negative exposure to the market factor.

```
crsp_monthly |>
  left_join(beta_monthly, by = c("permno", "month")) |>
  drop_na(beta_monthly) |>
  group_by(industry, permno) |>
  summarize(beta = mean(beta_monthly)) |>
  ggplot(aes(x = reorder(industry, beta, FUN = median), y = beta)) +
  geom_boxplot() +
  coord_flip() +
  labs(
    x = NULL, y = NULL,
    title = "Firm-specific beta distributions by industry"
  )
```

Next, we illustrate the time-variation in the cross-section of estimated betas. Figure 6.3 shows the monthly deciles of estimated betas (based on monthly data) and indicates an interesting pattern: First, betas seem to vary over time in the sense that during some periods, there is a clear trend across all deciles. Second, the sample exhibits periods where the dispersion across stocks increases in the sense that the lower decile decreases and the upper decile increases, which indicates that for some stocks the correlation with the market increases while for others it decreases. Note also here: stocks with negative betas are a rare exception.

```
beta_monthly |>
  drop_na(beta_monthly) |>
  group_by(month) |>
  summarize(
    x = quantile(beta_monthly, seq(0.1, 0.9, 0.1)),
    quantile = 100 * seq(0.1, 0.9, 0.1),
    .groups = "drop"
  ) |>
  ggplot(aes(
    x = month,
    y = x,
    color = as_factor(quantile),
    linetype = as_factor(quantile)
    )) +
  geom_line() +
  labs(
    x = NULL, y = NULL, color = NULL, linetype = NULL,
    title = "Monthly deciles of estimated betas",
  )
```

To compare the difference between daily and monthly data, we combine beta estimates to a single table. Then, we use the table to plot a comparison of beta estimates for our example stocks in Figure 6.4.

```
beta <- beta_monthly |>
  full_join(beta_daily, by = c("permno", "month")) |>
  arrange(permno, month)

beta |>
  inner_join(examples, by = "permno") |>
  pivot_longer(cols = c(beta_monthly, beta_daily)) |>
  drop_na() |>
  ggplot(aes(
    x = month,
    y = value,
    color = name,
    linetype = name
```

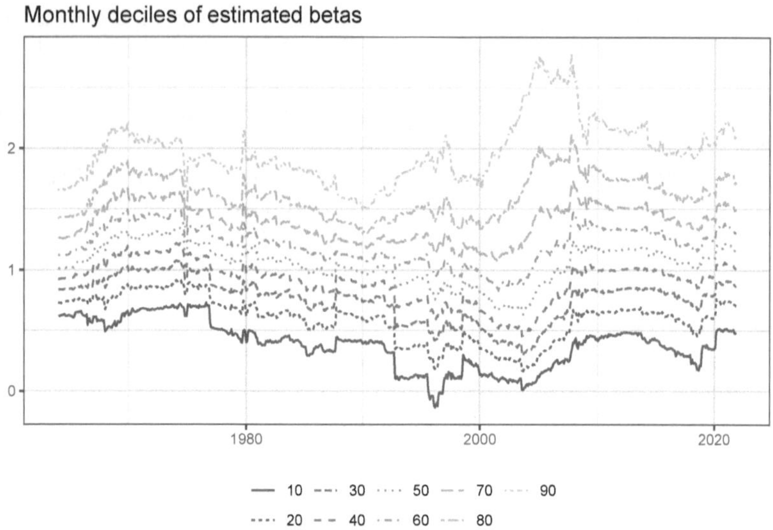

FIGURE 6.3
Each line corresponds to the monthly cross-sectional quantile of the estimated CAPM beta.

```
  )) +
  geom_line() +
  facet_wrap(~company, ncol = 1) +
  labs(
    x = NULL, y = NULL, color = NULL, linetype = NULL,
    title = "Comparison of beta estimates using monthly and daily data"
  )
```

The estimates in Figure 6.4 look as expected. As you can see, it really depends on the estimation window and data frequency how your beta estimates turn out.

Finally, we write the estimates to our database such that we can use them in later chapters.

```
dbWriteTable(tidy_finance,
  "beta",
  value = beta,
  overwrite = TRUE
)
```

Whenever you perform some kind of estimation, it also makes sense to do rough plausibility tests. A possible check is to plot the share of stocks with beta estimates

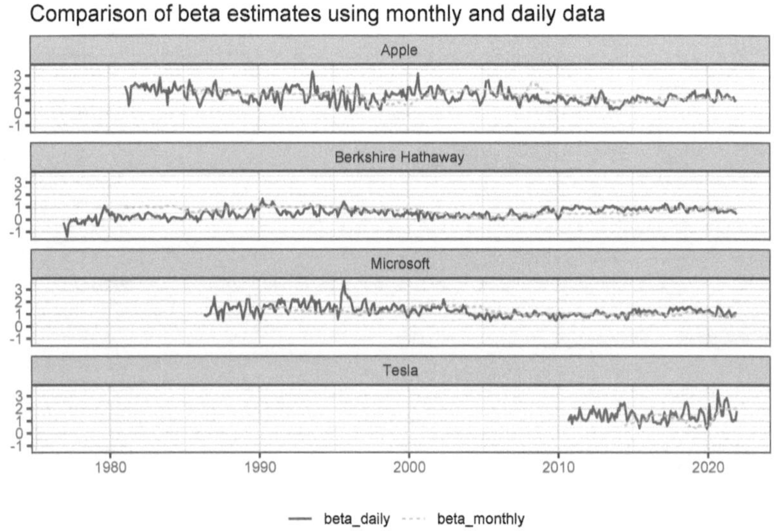

FIGURE 6.4
CAPM betas are computed using 5 years of monthly or 3 months of daily data. The two lines show the monthly estimates based on a rolling window for few exemplary stocks.

over time. This descriptive helps us discover potential errors in our data preparation or estimation procedure. For instance, suppose there was a gap in our output where we do not have any betas. In this case, we would have to go back and check all previous steps to find out what went wrong.

```
beta_long <- crsp_monthly |>
  left_join(beta, by = c("permno", "month")) |>
  pivot_longer(cols = c(beta_monthly, beta_daily))

beta_long |>
  group_by(month, name) |>
  summarize(share = sum(!is.na(value)) / n()) |>
  ggplot(aes(
    x = month,
    y = share,
    color = name,
    linetype = name
    )) +
  geom_line() +
  scale_y_continuous(labels = percent) +
  labs(
    x = NULL, y = NULL, color = NULL, linetype = NULL,
```

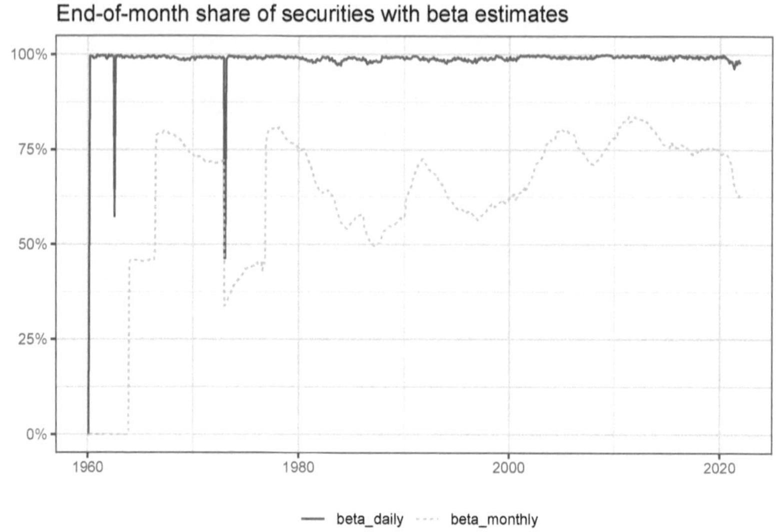

FIGURE 6.5
The two lines show the share of securities with beta estimates using 5 years of monthly or 3 months of daily data.

```
    title = "End-of-month share of securities with beta estimates"
  ) +
  coord_cartesian(ylim = c(0, 1))
```

Figure 6.5 does not indicate any troubles, so let us move on to the next check.

We also encourage everyone to always look at the distributional summary statistics of variables. You can easily spot outliers or weird distributions when looking at such tables.

```
beta_long |>
  select(name, value) |>
  drop_na() |>
  group_by(name) |>
  summarize(
    mean = mean(value),
    sd = sd(value),
    min = min(value),
    q05 = quantile(value, 0.05),
    q50 = quantile(value, 0.50),
    q95 = quantile(value, 0.95),
    max = max(value),
```

```
  n = n()
)
```

```
# A tibble: 2 x 9
  name          mean    sd   min    q05   q50   q95   max       n
  <chr>        <dbl> <dbl> <dbl>  <dbl> <dbl> <dbl> <dbl>   <int>
1 beta_daily   0.749 0.926 -43.7 -0.447 0.686  2.23  56.6 3233745
2 beta_monthly 1.10  0.713 -13.0  0.125 1.03   2.32  10.3 2102936
```

The summary statistics also look plausible for the two estimation procedures.

Finally, since we have two different estimators for the same theoretical object, we expect the estimators should be at least positively correlated (although not perfectly as the estimators are based on different sample periods and frequencies).

```
beta |>
  select(beta_daily, beta_monthly) |>
  cor(use = "complete.obs")
```

```
             beta_daily beta_monthly
beta_daily        1.000        0.323
beta_monthly      0.323        1.000
```

Indeed, we find a positive correlation between our beta estimates. In the subsequent chapters, we mainly use the estimates based on monthly data as most readers should be able to replicate them due to potential memory limitations that might arise with the daily data.

6.6 Exercises

1. Compute beta estimates based on monthly data using 1, 3, and 5 years of data and impose a minimum number of observations of 10, 28, and 48 months with return data, respectively. How strongly correlated are the estimated betas?
2. Compute beta estimates based on monthly data using 5 years of data and impose different numbers of minimum observations. How does the share of permno-month observations with successful beta estimates vary across the different requirements? Do you find a high correlation across the estimated betas?

3. Instead of using `future_map()`, perform the beta estimation in a loop (using either monthly or daily data) for a subset of 100 permnos of your choice. Verify that you get the same results as with the parallelized code from above.
4. Filter out the stocks with negative betas. Do these stocks frequently exhibit negative betas, or do they resemble estimation errors?
5. Compute beta estimates for multi-factor models such as the Fama-French 3 factor model. For that purpose, you extend your regression to

$$r_{i,t} - r_{f,t} = \alpha_i + \sum_{j=1}^{k} \beta_{i,k}(r_{j,t} - r_{f,t}) + \varepsilon_{i,t}$$

where $r_{j,t}$ are the k factor returns. Thus, you estimate 4 parameters (α_i and the slope coefficients). Provide some summary statistics of the cross-section of firms and their exposure to the different factors.

7

Univariate Portfolio Sorts

In this chapter, we dive into portfolio sorts, one of the most widely used statistical methodologies in empirical asset pricing (e.g., Bali et al., 2016). The key application of portfolio sorts is to examine whether one or more variables can predict future excess returns. In general, the idea is to sort individual stocks into portfolios, where the stocks within each portfolio are similar with respect to a sorting variable, such as firm size. The different portfolios then represent well-diversified investments that differ in the level of the sorting variable. You can then attribute the differences in the return distribution to the impact of the sorting variable. We start by introducing univariate portfolio sorts (which sort based on only one characteristic) and tackle bivariate sorting in Chapter 9.

A univariate portfolio sort considers only one sorting variable $x_{t-1,i}$. Here, i denotes the stock and $t-1$ indicates that the characteristic is observable by investors at time t.
The objective is to assess the cross-sectional relation between $x_{t-1,i}$ and, typically, stock excess returns $r_{t,i}$ at time t as the outcome variable. To illustrate how portfolio sorts work, we use estimates for market betas from the previous chapter as our sorting variable.

The current chapter relies on the following set of packages.

```
library(tidyverse)
library(RSQLite)
library(lubridate)
library(scales)
library(lmtest)
library(sandwich)
```

Compared to previous chapters, we introduce `lmtest` (Zeileis and Hothorn, 2002) for inference for estimated coefficients, and `sandwich` (Zeileis, 2006) for different covariance matrix estimators.

DOI: 10.1201/b23237-7

7.1 Data Preparation

We start with loading the required data from our `SQLite`-database introduced in Chapters 2–4. In particular, we use the monthly CRSP sample as our asset universe. Once we form our portfolios, we use the Fama-French market factor returns to compute the risk-adjusted performance (i.e., alpha). `beta` is the tibble with market betas computed in the previous chapter.

```
tidy_finance <- dbConnect(
  SQLite(),
  "data/tidy_finance.sqlite",
  extended_types = TRUE
)

crsp_monthly <- tbl(tidy_finance, "crsp_monthly") |>
  select(permno, month, ret_excess, mktcap_lag) |>
  collect()

factors_ff_monthly <- tbl(tidy_finance, "factors_ff_monthly") |>
  collect()

beta <- tbl(tidy_finance, "beta") |>
  collect()
```

7.2 Sorting by Market Beta

Next, we merge our sorting variable with the return data. We use the one-month *lagged* betas as a sorting variable to ensure that the sorts rely only on information available when we create the portfolios. To lag stock beta by one month, we add one month to the current date and join the resulting information with our return data. This procedure ensures that month t information is available in month $t + 1$. You may be tempted to simply use a call such as `crsp_monthly |> group_by(permno) |> mutate(beta_lag = lag(beta)))` instead. This procedure, however, does not work correctly if there are non-explicit missing values in the time series.

```
beta_lag <- beta |>
  mutate(month = month %m+% months(1)) |>
  select(permno, month, beta_lag = beta_monthly) |>
  drop_na()
```

```
data_for_sorts <- crsp_monthly |>
  inner_join(beta_lag, by = c("permno", "month"))
```

The first step to conduct portfolio sorts is to calculate periodic breakpoints that you can use to group the stocks into portfolios. For simplicity, we start with the median lagged market beta as the single breakpoint. We then compute the value-weighted returns for each of the two resulting portfolios, which means that the lagged market capitalization determines the weight in `weighted.mean()`.

```
beta_portfolios <- data_for_sorts |>
  group_by(month) |>
  mutate(
    breakpoint = median(beta_lag),
    portfolio = case_when(
      beta_lag <= breakpoint ~ "low",
      beta_lag > breakpoint ~ "high"
    )
  ) |>
  group_by(month, portfolio) |>
  summarize(ret = weighted.mean(ret_excess, mktcap_lag), .groups = "drop")
```

7.3 Performance Evaluation

We can construct a long-short strategy based on the two portfolios: buy the high-beta portfolio and, at the same time, short the low-beta portfolio. Thereby, the overall position in the market is net-zero, i.e., you do not need to invest money to realize this strategy in the absence of frictions.

```
beta_longshort <- beta_portfolios |>
  pivot_wider(month, names_from = portfolio, values_from = ret) |>
  mutate(long_short = high - low)
```

We compute the average return and the corresponding standard error to test whether the long-short portfolio yields on average positive or negative excess returns. In the asset pricing literature, one typically adjusts for autocorrelation by using Newey and West (1987) t-statistics to test the null hypothesis that average portfolio excess returns are equal to zero. One necessary input for Newey-West standard errors is a chosen bandwidth based on the number of lags employed for the estimation. While it seems that researchers often default on choosing a pre-specified lag length of 6 months, we instead recommend a data-driven approach. This automatic selection is advocated by Newey and West (1994) and available in the `sandwich` package. To implement this test, we compute the average return via `lm()` and then employ the `coeftest()` function. If you want to implement the typical 6-lag default setting, you can enforce it by passing the arguments `lag = 6, prewhite = FALSE` to the `coeftest()` function in the code below and it passes them on to `NeweyWest()`.

```
model_fit <- lm(long_short ~ 1, data = beta_longshort)
coeftest(model_fit, vcov = NeweyWest)
```

```
t test of coefficients:

            Estimate Std. Error t value Pr(>|t|)
(Intercept) 0.000287   0.001312    0.22     0.83
```

The results indicate that we cannot reject the null hypothesis of average returns being equal to zero. Our portfolio strategy using the median as a breakpoint hence does not yield any abnormal returns. Is this finding surprising if you reconsider the CAPM? It certainly is. The CAPM yields that the high beta stocks should yield higher expected returns. Our portfolio sort implicitly mimics an investment strategy that finances high beta stocks by shorting low beta stocks. Therefore, one should expect that the average excess returns yield a return that is above the risk-free rate.

7.4 Functional Programming for Portfolio Sorts

Now we take portfolio sorts to the next level. We want to be able to sort stocks into an arbitrary number of portfolios. For this case, functional programming is very handy: we employ the curly-curly[1]-operator to give us flexibility concerning which variable to use for the sorting, denoted by `var`. We use `quantile()` to compute breakpoints for `n_portfolios`. Then, we assign portfolios to stocks using the `findInterval()` function. The output of the following function is a new column that contains the number of the portfolio to which a stock belongs.

```
assign_portfolio <- function(data, var, n_portfolios) {
  breakpoints <- data |>
    summarize(breakpoint = quantile({{ var }},
      probs = seq(0, 1, length.out = n_portfolios + 1),
      na.rm = TRUE
    )) |>
```

[1]https://www.tidyverse.org/blog/2019/06/rlang-0-4-0/#a-simpler-interpolation-pattern-with-

```
    pull(breakpoint) |>
    as.numeric()

  assigned_portfolios <- data |>
    mutate(portfolio = findInterval({{ var }},
      breakpoints,
      all.inside = TRUE
    )) |>
    pull(portfolio)

  return(assigned_portfolios)
}
```

We can use the above function to sort stocks into ten portfolios each month using lagged betas and compute value-weighted returns for each portfolio. Note that we transform the portfolio column to a factor variable because it provides more convenience for the figure construction below.

```
beta_portfolios <- data_for_sorts |>
  group_by(month) |>
  mutate(
    portfolio = assign_portfolio(
      data = cur_data(),
      var = beta_lag,
      n_portfolios = 10
    ),
    portfolio = as.factor(portfolio)
  ) |>
  group_by(portfolio, month) |>
  summarize(
    ret = weighted.mean(ret_excess, mktcap_lag),
    .groups = "drop"
  )
```

7.5 More Performance Evaluation

In the next step, we compute summary statistics for each beta portfolio. Namely, we compute CAPM-adjusted alphas, the beta of each beta portfolio, and average returns.

```
beta_portfolios_summary <- beta_portfolios |>
  left_join(factors_ff_monthly, by = "month") |>
  group_by(portfolio) |>
  summarize(
    alpha = as.numeric(lm(ret ~ 1 + mkt_excess)$coefficients[1]),
    beta = as.numeric(lm(ret ~ 1 + mkt_excess)$coefficients[2]),
    ret = mean(ret)
  )
```

Figure 7.1 illustrates the CAPM alphas of beta-sorted portfolios. It shows that low beta portfolios tend to exhibit positive alphas, while high beta portfolios exhibit negative alphas.

```
beta_portfolios_summary |>
  ggplot(aes(x = portfolio, y = alpha, fill = portfolio)) +
  geom_bar(stat = "identity") +
  labs(
    title = "CAPM alphas of beta-sorted portfolios",
    x = "Portfolio",
    y = "CAPM alpha",
    fill = "Portfolio"
  ) +
  scale_y_continuous(labels = percent) +
  theme(legend.position = "None")
```

These results suggest a negative relation between beta and future stock returns, which contradicts the predictions of the CAPM. According to the CAPM, returns should increase with beta across the portfolios and risk-adjusted returns should be statistically indistinguishable from zero.

7.6 The Security Market Line and Beta Portfolios

The CAPM predicts that our portfolios should lie on the security market line (SML). The slope of the SML is equal to the market risk premium and reflects the risk-return trade-off at any given time. Figure 7.2 illustrates the security market line: We see that (not surprisingly) the high beta portfolio returns have a high correlation with the market returns. However, it seems like the average excess returns for high beta stocks are lower than what the security market line implies would be an "appropriate" compensation for the high market risk.

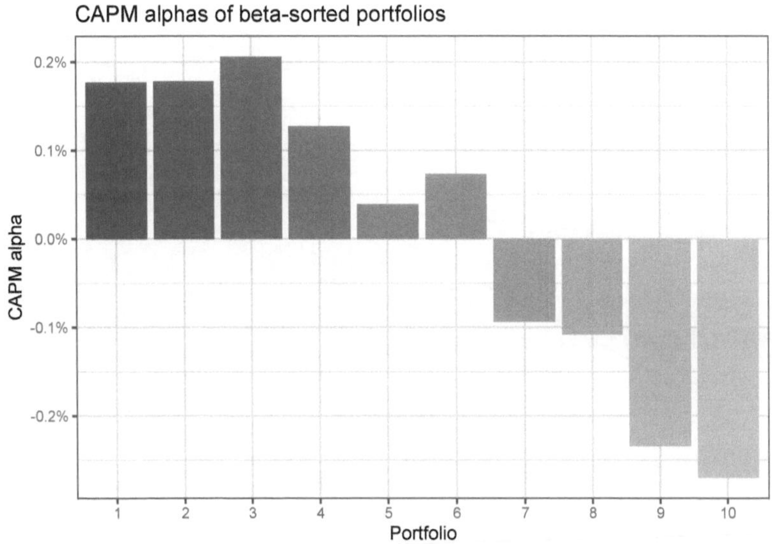

FIGURE 7.1
Portfolios are sorted into deciles each month based on their estimated CAPM beta. The bar charts indicate the CAPM alpha of the resulting portfolio returns during the entire CRSP period.

```
sml_capm <- lm(ret ~ 1 + beta, data = beta_portfolios_summary)$coefficients

beta_portfolios_summary |>
  ggplot(aes(
    x = beta,
    y = ret,
    color = portfolio
  )) +
  geom_point() +
  geom_abline(
    intercept = 0,
    slope = mean(factors_ff_monthly$mkt_excess),
    linetype = "solid"
  ) +
  geom_abline(
    intercept = sml_capm[1],
    slope = sml_capm[2],
    linetype = "dashed"
  ) +
  scale_y_continuous(
    labels = percent,
    limit = c(0, mean(factors_ff_monthly$mkt_excess) * 2)
```

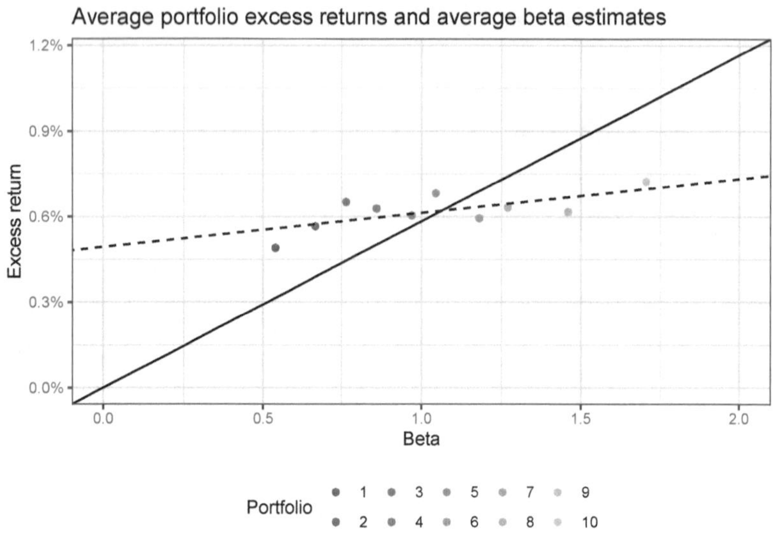

FIGURE 7.2
Excess returns are computed as CAPM alphas of the beta-sorted portfolios. The horizontal axis indicates the CAPM beta of the resulting beta-sorted portfolio return time series. The dashed line indicates the slope coefficient of a linear regression of excess returns on portfolio betas.

```
  ) +
  scale_x_continuous(limits = c(0, 2)) +
  labs(
    x = "Beta", y = "Excess return", color = "Portfolio",
    title = "Average portfolio excess returns and average beta estimates"
  )
```

To provide more evidence against the CAPM predictions, we again form a long-short strategy that buys the high-beta portfolio and shorts the low-beta portfolio.

```
beta_longshort <- beta_portfolios |>
  ungroup() |>
  mutate(portfolio = case_when(
    portfolio == max(as.numeric(portfolio)) ~ "high",
    portfolio == min(as.numeric(portfolio)) ~ "low"
  )) |>
  filter(portfolio %in% c("low", "high")) |>
  pivot_wider(month, names_from = portfolio, values_from = ret) |>
  mutate(long_short = high - low) |>
  left_join(factors_ff_monthly, by = "month")
```

Again, the resulting long-short strategy does not exhibit statistically significant returns.

```
coeftest(lm(long_short ~ 1, data = beta_longshort),
  vcov = NeweyWest
)
```

```
t test of coefficients:

            Estimate Std. Error t value Pr(>|t|)
(Intercept)  0.00232    0.00322    0.72     0.47
```

However, the long-short portfolio yields a statistically significant negative CAPM-adjusted alpha, although, controlling for the effect of beta, the average excess stock returns should be zero according to the CAPM. The results thus provide no evidence in support of the CAPM. The negative value has been documented as the so-called betting against beta factor (Frazzini and Pedersen, 2014). Betting against beta corresponds to a strategy that shorts high beta stocks and takes a (levered) long position in low beta stocks. If borrowing constraints prevent investors from taking positions on the SML they are instead incentivized to buy high beta stocks, which leads to a relatively higher price (and therefore lower expected returns than implied by the CAPM) for such high beta stocks. As a result, the betting-against-beta strategy earns from providing liquidity to capital constraint investors with lower risk aversion.

```
coeftest(lm(long_short ~ 1 + mkt_excess, data = beta_longshort),
  vcov = NeweyWest
)
```

```
t test of coefficients:

            Estimate Std. Error t value Pr(>|t|)
(Intercept) -0.00447    0.00256   -1.75    0.081 .
mkt_excess   1.16555    0.09562   12.19   <2e-16 ***
---
Signif. codes:  0 '***' 0.001 '**' 0.01 '*' 0.05 '.' 0.1 ' ' 1
```

Figure 7.3 shows the annual returns of the extreme beta portfolios we are mainly interested in. The figure illustrates no consistent striking patterns over the last years – each portfolio exhibits periods with positive and negative annual returns.

```
beta_longshort |>
  group_by(year = year(month)) |>
  summarize(
    low = prod(1 + low),
    high = prod(1 + high),
```

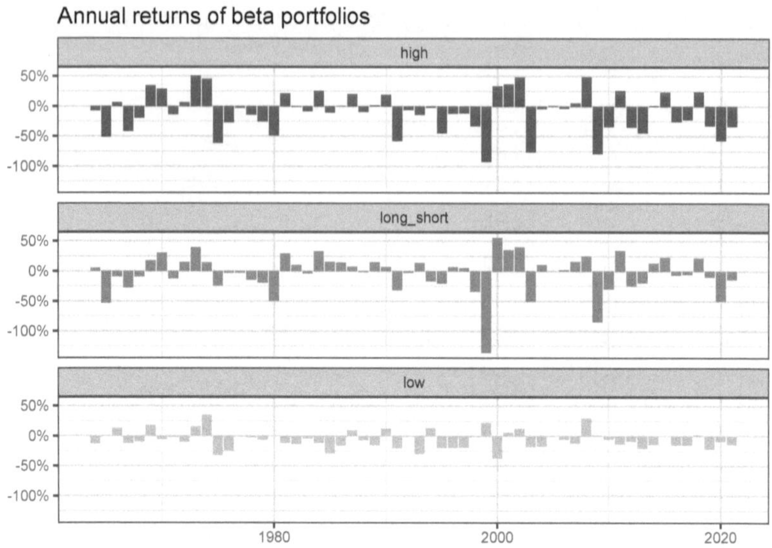

FIGURE 7.3
We construct portfolios by sorting stocks into high and low based on their estimated CAPM beta. Long short indicates a strategy that goes long into high beta stocks and short low beta stocks.

```
    long_short = prod(1 + long_short)
  ) |>
  pivot_longer(cols = -year) |>
  ggplot(aes(x = year, y = 1 - value, fill = name)) +
  geom_col(position = "dodge") +
  facet_wrap(~name, ncol = 1) +
  theme(legend.position = "none") +
  scale_y_continuous(labels = percent) +
  labs(
    title = "Annual returns of beta portfolios",
    x = NULL, y = NULL
  )
```

Overall, this chapter shows how functional programming can be leveraged to form an arbitrary number of portfolios using any sorting variable and how to evaluate the performance of the resulting portfolios. In the next chapter, we dive deeper into the many degrees of freedom that arise in the context of portfolio analysis.

7.7 Exercises

1. Take the two long-short beta strategies based on different numbers of portfolios and compare the returns. Is there a significant difference in returns? How do the Sharpe ratios compare between the strategies? Find one additional portfolio evaluation statistic and compute it.
2. We plotted the alphas of the ten beta portfolios above. Write a function that tests these estimates for significance. Which portfolios have significant alphas?
3. The analysis here is based on betas from monthly returns. However, we also computed betas from daily returns. Re-run the analysis and point out differences in the results.
4. Given the results in this chapter, can you define a long-short strategy that yields positive abnormal returns (i.e., alphas)? Plot the cumulative excess return of your strategy and the market excess return for comparison.

8

Size Sorts and p-Hacking

In this chapter, we continue with portfolio sorts in a univariate setting. Yet, we consider firm size as a sorting variable, which gives rise to a well-known return factor: the size premium. The size premium arises from buying small stocks and selling large stocks. Prominently, Fama and French (1993) include it as a factor in their three-factor model. Apart from that, asset managers commonly include size as a key firm characteristic when making investment decisions.

We also introduce new choices in the formation of portfolios. In particular, we discuss listing exchanges, industries, weighting regimes, and periods. These choices matter for the portfolio returns and result in different size premiums (see Hasler, 2021, Soebhag et al. (2022), and Walter et al. (2022) for more insights into decision nodes and their effect on premiums). Exploiting these ideas to generate favorable results is called p-hacking. There is arguably a thin line between p-hacking and conducting robustness tests. Our purpose here is to illustrate the substantial variation that can arise along the evidence-generating process.

The chapter relies on the following set of packages:

```
library(tidyverse)
library(RSQLite)
library(lubridate)
library(scales)
library(sandwich)
library(lmtest)
library(furrr)
library(rlang)
```

Compared to previous chapters, we introduce the `rlang` package (Henry and Wickham, 2022) for more advanced parsing of functional expressions.

DOI: 10.1201/b23237-8

8.1 Data Preparation

First, we retrieve the relevant data from our `SQLite`-database introduced in Chapters 2–4. Firm size is defined as market equity in most asset pricing applications that we retrieve from CRSP. We further use the Fama-French factor returns for performance evaluation.

```
tidy_finance <- dbConnect(
  SQLite(),
  "data/tidy_finance.sqlite",
  extended_types = TRUE
)

crsp_monthly <- tbl(tidy_finance, "crsp_monthly") |>
  collect()

factors_ff_monthly <- tbl(tidy_finance, "factors_ff_monthly") |>
  collect()
```

8.2 Size Distribution

Before we build our size portfolios, we investigate the distribution of the variable *firm size*. Visualizing the data is a valuable starting point to understand the input to the analysis. Figure 8.1 shows the fraction of total market capitalization concentrated in the largest firm. To produce this graph, we create monthly indicators that track whether a stock belongs to the largest x percent of the firms. Then, we aggregate the firms within each bucket and compute the buckets' share of total market capitalization.

Figure 8.1 shows that the largest 1 percent of firms cover up to 50 percent of the total market capitalization, and holding just the 25 percent largest firms in the CRSP universe essentially replicates the market portfolio. The distribution of firm size thus implies that the largest firms of the market dominate many small firms whenever we use value-weighted benchmarks.

```
crsp_monthly |>
  group_by(month) |>
  mutate(
    top01 = if_else(mktcap >= quantile(mktcap, 0.99), 1, 0),
```

```
    top05 = if_else(mktcap >= quantile(mktcap, 0.95), 1, 0),
    top10 = if_else(mktcap >= quantile(mktcap, 0.90), 1, 0),
    top25 = if_else(mktcap >= quantile(mktcap, 0.75), 1, 0),
    total_market_cap = sum(mktcap)
  ) |>
  summarize(
    `Largest 1% of stocks` = sum(mktcap[top01 == 1]) / total_market_cap,
    `Largest 5% of stocks` = sum(mktcap[top05 == 1]) / total_market_cap,
    `Largest 10% of stocks` = sum(mktcap[top10 == 1]) / total_market_cap,
    `Largest 25% of stocks` = sum(mktcap[top25 == 1]) / total_market_cap
  ) |>
  pivot_longer(cols = -month) |>
  mutate(name = factor(name, levels = c(
    "Largest 1% of stocks", "Largest 5% of stocks",
    "Largest 10% of stocks", "Largest 25% of stocks"
  ))) |>
  ggplot(aes(
    x = month,
    y = value,
    color = name,
    linetype = name)) +
  geom_line() +
  scale_y_continuous(labels = percent, limits = c(0, 1)) +
  labs(
    x = NULL, y = NULL, color = NULL, linetype = NULL,
    title = "Percentage of total market capitalization in largest stocks"
  )
```

Next, firm sizes also differ across listing exchanges. Stocks' primary listings were important in the past and are potentially still relevant today. Figure 8.2 shows that the New York Stock Exchange (NYSE) was and still is the largest listing exchange in terms of market capitalization. More recently, NASDAQ has gained relevance as a listing exchange. Do you know what the small peak in NASDAQ's market cap around the year 2000 was?

```
crsp_monthly |>
  group_by(month, exchange) |>
  summarize(mktcap = sum(mktcap)) |>
  mutate(share = mktcap / sum(mktcap)) |>
  ggplot(aes(
    x = month,
    y = share,
    fill = exchange,
    color = exchange)) +
  geom_area(
```

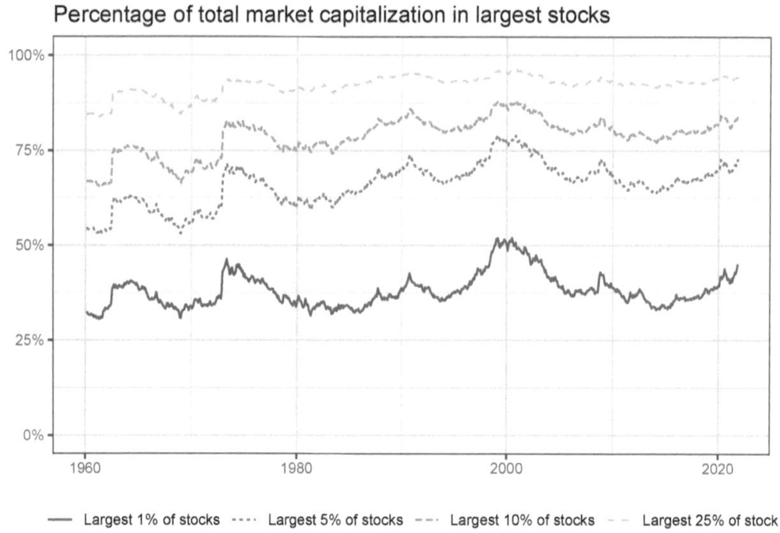

FIGURE 8.1
We report the aggregate market capitalization of all stocks that belong to the 1, 5, 10, and 25 percent quantile of the largest firms in the monthly cross-section relative to the market capitalization of all stocks during the month.

```
    position = "stack",
    stat = "identity",
    alpha = 0.5
  ) +
  geom_line(position = "stack") +
  scale_y_continuous(labels = percent) +
  labs(
    x = NULL, y = NULL, fill = NULL, color = NULL,
    title = "Share of total market capitalization per listing exchange"
  )
```

Finally, we consider the distribution of firm size across listing exchanges and create summary statistics. The function `summary()` does not include all statistics we are interested in, which is why we create the function `create_summary()` that adds the standard deviation and the number of observations. Then, we apply it to the most current month of our CRSP data on each listing exchange. We also add a row with `add_row()` with the overall summary statistics.

The resulting table shows that firms listed on NYSE in December 2021 are significantly larger on average than firms listed on the other exchanges. Moreover, NASDAQ lists the largest number of firms. This discrepancy between firm sizes across listing exchanges motivated researchers to form breakpoints exclusively on the NYSE sample

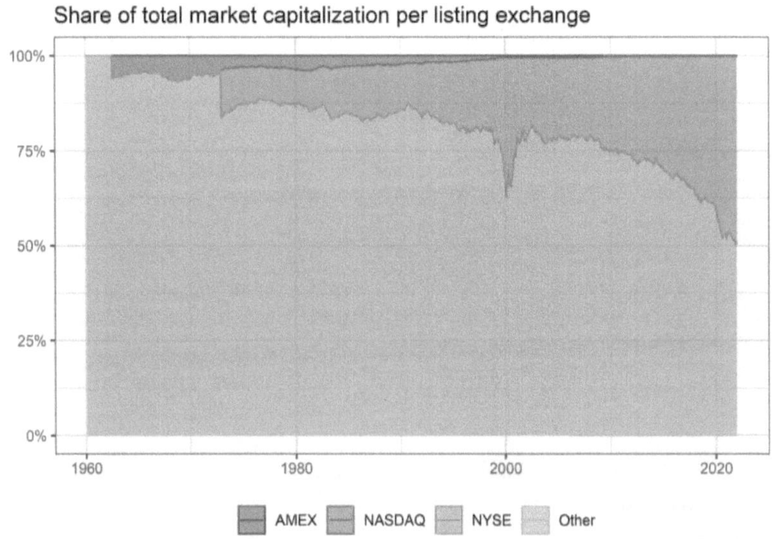

FIGURE 8.2
Years are on the horizontal axis and the corresponding share of total market capitalization per listing exchange on the vertical axis.

and apply those breakpoints to all stocks. In the following, we use this distinction to update our portfolio sort procedure.

```
create_summary <- function(data, column_name) {
  data |>
    select(value = {{ column_name }}) |>
    summarize(
      mean = mean(value),
      sd = sd(value),
      min = min(value),
      q05 = quantile(value, 0.05),
      q50 = quantile(value, 0.50),
      q95 = quantile(value, 0.95),
      max = max(value),
      n = n()
    )
}

crsp_monthly |>
  filter(month == max(month)) |>
  group_by(exchange) |>
  create_summary(mktcap) |>
  add_row(crsp_monthly |>
```

```
    filter(month == max(month)) |>
    create_summary(mktcap) |>
    mutate(exchange = "Overall"))
```

```
# A tibble: 5 x 9
  exchange    mean     sd      min     q05     q50    q95     max     n
  <chr>      <dbl>  <dbl>    <dbl>   <dbl>   <dbl>  <dbl>   <dbl> <int>
1 AMEX        415.  2181.     7.57    12.6    75.8  1218. 2.57e4   145
2 NASDAQ     8649. 90038.     7.01    29.1   428.  18781. 2.90e6  2779
3 NYSE      17858. 48619.    23.9    195.   3434.  80748. 4.73e5  1395
4 Other     13906.     NA 13906.  13906.  13906.  13906. 1.39e4     1
5 Overall   11348. 77458.     7.01    34.0   794.  40647. 2.90e6  4320
```

8.3 Univariate Size Portfolios with Flexible Breakpoints

In Chapter 7, we construct portfolios with a varying number of breakpoints and different sorting variables. Here, we extend the framework such that we compute breakpoints on a subset of the data, for instance, based on selected listing exchanges. In published asset pricing articles, many scholars compute sorting breakpoints only on NYSE-listed stocks. These NYSE-specific breakpoints are then applied to the entire universe of stocks.

To replicate the NYSE-centered sorting procedure, we introduce `exchanges` as an argument in our `assign_portfolio()` function. The exchange-specific argument then enters in the filter `filter(exchange %in% exchanges)`. For example, if `exchanges = 'NYSE'` is specified, only stocks listed on NYSE are used to compute the breakpoints. Alternatively, you could specify `exchanges = c("NYSE", "NASDAQ", "AMEX")`, which keeps all stocks listed on either of these exchanges. Overall, regular expressions are a powerful tool, and we only touch on a specific case here.

```
assign_portfolio <- function(n_portfolios,
                             exchanges,
                             data) {
  breakpoints <- data |>
    filter(exchange %in% exchanges) |>
    summarize(breakpoint = quantile(
      mktcap_lag,
      probs = seq(0, 1, length.out = n_portfolios + 1),
      na.rm = TRUE
    )) |>
    pull(breakpoint) |>
```

```
    as.numeric()

  assigned_portfolios <- data |>
    mutate(portfolio = findInterval(mktcap_lag,
      breakpoints,
      all.inside = TRUE
    )) |>
    pull(portfolio)
  return(assigned_portfolios)
}
```

8.4 Weighting Schemes for Portfolios

Apart from computing breakpoints on different samples, researchers often use different portfolio weighting schemes. So far, we weighted each portfolio constituent by its relative market equity of the previous period. This protocol is called *value-weighting*. The alternative protocol is *equal-weighting*, which assigns each stock's return the same weight, i.e., a simple average of the constituents' returns. Notice that equal-weighting is difficult in practice as the portfolio manager needs to rebalance the portfolio monthly while value-weighting is a truly passive investment.

We implement the two weighting schemes in the function `compute_portfolio_returns()` that takes a logical argument to weight the returns by firm value. The statement `if_else(value_weighted, weighted.mean(ret_excess, mktcap_lag), mean(ret_excess))` generates value-weighted returns if `value_weighted = TRUE`. Additionally, the long-short portfolio is long in the smallest firms and short in the largest firms, consistent with research showing that small firms outperform their larger counterparts. Apart from these two changes, the function is similar to the procedure in Chapter 7.

```
compute_portfolio_returns <- function(n_portfolios = 10,
                                      exchanges = c("NYSE", "NASDAQ", "AMEX"),
                                      value_weighted = TRUE,
                                      data = crsp_monthly) {
  data |>
    group_by(month) |>
    mutate(portfolio = assign_portfolio(
      n_portfolios = n_portfolios,
      exchanges = exchanges,
      data = cur_data()
    )) |>
```

```
    group_by(month, portfolio) |>
    summarize(
      ret = if_else(value_weighted,
        weighted.mean(ret_excess, mktcap_lag),
        mean(ret_excess)
      ),
      .groups = "drop_last"
    ) |>
    summarize(size_premium = ret[portfolio == min(portfolio)] -
      ret[portfolio == max(portfolio)]) |>
    summarize(size_premium = mean(size_premium))
}
```

To see how the function `compute_portfolio_returns()` works, we consider a simple median breakpoint example with value-weighted returns. We are interested in the effect of restricting listing exchanges on the estimation of the size premium. In the first function call, we compute returns based on breakpoints from all listing exchanges. Then, we computed returns based on breakpoints from NYSE-listed stocks.

```
ret_all <- compute_portfolio_returns(
  n_portfolios = 2,
  exchanges = c("NYSE", "NASDAQ", "AMEX"),
  value_weighted = TRUE,
  data = crsp_monthly
)

ret_nyse <- compute_portfolio_returns(
  n_portfolios = 2,
  exchanges = "NYSE",
  value_weighted = TRUE,
  data = crsp_monthly
)

tibble(
  Exchanges = c("NYSE, NASDAQ & AMEX", "NYSE"),
  Premium = as.numeric(c(ret_all, ret_nyse)) * 100
)
```

```
# A tibble: 2 x 2
  Exchanges           Premium
  <chr>                 <dbl>
1 NYSE, NASDAQ & AMEX  0.0975
2 NYSE                 0.166
```

The table shows that the size premium is more than 60 percent larger if we consider only stocks from NYSE to form the breakpoint each month. The NYSE-specific breakpoints are larger, and there are more than 50 percent of the stocks in the entire universe in the resulting small portfolio because NYSE firms are larger on average. The impact of this choice is not negligible.

8.5 P-hacking and Non-standard Errors

Since the choice of the listing exchange has a significant impact, the next step is to investigate the effect of other data processing decisions researchers have to make along the way. In particular, any portfolio sort analysis has to decide at least on the number of portfolios, the listing exchanges to form breakpoints, and equal- or value-weighting. Further, one may exclude firms that are active in the finance industry or restrict the analysis to some parts of the time series. All of the variations of these choices that we discuss here are part of scholarly articles published in the top finance journals. We refer to Walter et al. (2022) for an extensive set of other decision nodes at the discretion of researchers.

The intention of this application is to show that the different ways to form portfolios result in different estimated size premiums. Despite the effects of this multitude of choices, there is no correct way. It should also be noted that none of the procedures is wrong, the aim is simply to illustrate the changes that can arise due to the variation in the evidence-generating process (Menkveld et al., 2021). The term *non-standard errors* refers to the variation due to (suitable) choices made by researchers. Interestingly, in a large scale study, Menkveld et al. (2021) find that the magnitude of non-standard errors are similar than the estimation uncertainty based on a chosen model which shows how important it is to adjust for the seemingly innocent choices in the data preparation and evaluation workflow.

From a malicious perspective, these modeling choices give the researcher multiple *chances* to find statistically significant results. Yet this is considered *p-hacking*, which renders the statistical inference due to multiple testing invalid (Harvey et al., 2016).

Nevertheless, the multitude of options creates a problem since there is no single correct way of sorting portfolios. How should a researcher convince a reader that their results do not come from a p-hacking exercise? To circumvent this dilemma, academics are encouraged to present evidence from different sorting schemes as *robustness tests* and report multiple approaches to show that a result does not depend on a single choice. Thus, the robustness of premiums is a key feature.

Below we conduct a series of robustness tests which could also be interpreted as a p-hacking exercise. To do so, we examine the size premium in different specifications presented in the table `p_hacking_setup`. The function `expand_grid()` produces a table of all possible permutations of its arguments. Note that we use the argument `data` to exclude financial firms and truncate the time series.

```
p_hacking_setup <- expand_grid(
  n_portfolios = c(2, 5, 10),
  exchanges = list("NYSE", c("NYSE", "NASDAQ", "AMEX")),
  value_weighted = c(TRUE, FALSE),
  data = parse_exprs(
    'crsp_monthly;
     crsp_monthly |> filter(industry != "Finance");
     crsp_monthly |> filter(month < "1990-06-01");
     crsp_monthly |> filter(month >="1990-06-01")'
  )
)
```

To speed the computation up we parallelize the (many) different sorting procedures, as in the beta estimation of Chapter 6. Finally, we report the resulting size premiums in descending order. There are indeed substantial size premiums possible in our data, in particular when we use equal-weighted portfolios.

```
plan(multisession, workers = availableCores())

p_hacking_setup <- p_hacking_setup |>
  mutate(size_premium = future_pmap(
    .l = list(
      n_portfolios,
      exchanges,
      value_weighted,
      data
    ),
    .f = ~ compute_portfolio_returns(
      n_portfolios = ..1,
      exchanges = ..2,
      value_weighted = ..3,
      data = eval_tidy(..4)
    )
  ))

p_hacking_results <- p_hacking_setup |>
  mutate(data = map_chr(data, deparse)) |>
  unnest(size_premium) |>
  arrange(desc(size_premium))
p_hacking_results
```

```
# A tibble: 48 x 5
```

```
  n_portfolios exchanges value_weighted data                       size_~1
         <dbl> <list>    <lgl>          <chr>                        <dbl>
1           10 <chr [3]> FALSE          "filter(crsp_monthly~  0.0186
2           10 <chr [3]> FALSE          "filter(crsp_monthly~  0.0182
3           10 <chr [3]> FALSE          "crsp_monthly"         0.0163
4           10 <chr [3]> FALSE          "filter(crsp_monthly~  0.0139
5           10 <chr [3]> TRUE           "filter(crsp_monthly~  0.0115
# ... with 43 more rows, and abbreviated variable name
#   1: size_premium
```

8.6 The Size-Premium Variation

We provide a graph in Figure 8.3 that shows the different premiums. Figure 8.3 also shows the relation to the average Fama-French SMB (small minus big) premium used in the literature which we include as a dotted vertical line.

```
p_hacking_results |>
  ggplot(aes(x = size_premium)) +
  geom_histogram(bins = nrow(p_hacking_results)) +
  labs(
    x = NULL, y = NULL,
    title = "Distribution of size premiums for different sorting choices"
  ) +
  geom_vline(aes(xintercept = mean(factors_ff_monthly$smb)),
    linetype = "dashed"
  ) +
  scale_x_continuous(labels = percent)
```

8.7 Exercises

1. We gained several insights on the size distribution above. However, we did not analyze the average size across listing exchanges and industries. Which listing exchanges/industries have the largest firms? Plot the average firm size for the three listing exchanges over time. What do you conclude?
2. We compute breakpoints but do not take a look at them in the exposition above. This might cover potential data errors. Plot the breakpoints for

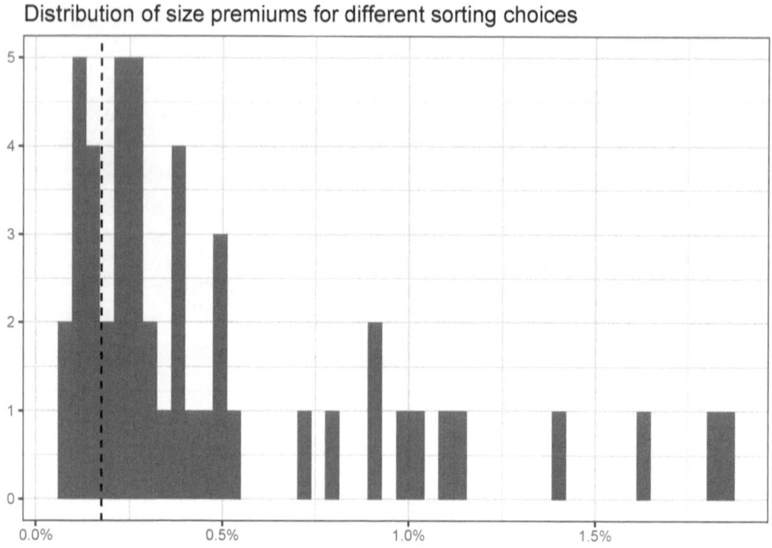

FIGURE 8.3
The dashed vertical line indicates the average Fama-French SMB premium.

ten size portfolios over time. Then, take the difference between the two extreme portfolios and plot it. Describe your results.

3. The returns that we analyse above do not account for differences in the exposure to market risk, i.e., the CAPM beta. Change the function `compute_portfolio_returns()` to output the CAPM alpha or beta instead of the average excess return.
4. While you saw the spread in returns from the p-hacking exercise, we did not show which choices led to the largest effects. Find a way to investigate which choice variable has the largest impact on the estimated size premium.
5. We computed several size premiums, but they do not follow the definition of Fama and French (1993). Which of our approaches comes closest to their SMB premium?

9

Value and Bivariate Sorts

In this chapter, we extend univariate portfolio analysis to bivariate sorts, which means we assign stocks to portfolios based on two characteristics. Bivariate sorts are regularly used in the academic asset pricing literature and are the basis for the Fama and French three factors. However, some scholars also use sorts with three grouping variables. Conceptually, portfolio sorts are easily applicable in higher dimensions.

We form portfolios on firm size and the book-to-market ratio. To calculate book-to-market ratios, accounting data is required, which necessitates additional steps during portfolio formation. In the end, we demonstrate how to form portfolios on two sorting variables using so-called independent and dependent portfolio sorts.

The current chapter relies on this set of packages.

```
library(tidyverse)
library(RSQLite)
library(lubridate)
library(scales)
library(lmtest)
library(sandwich)
```

9.1 Data Preparation

First, we load the necessary data from our SQLite-database introduced in Chapters 2–4. We conduct portfolio sorts based on the CRSP sample but keep only the necessary columns in our memory. We use the same data sources for firm size as in Chapter 8.

```
tidy_finance <- dbConnect(
  SQLite(),
  "data/tidy_finance.sqlite",
  extended_types = TRUE
)
```

DOI: 10.1201/b23237-9

```
crsp_monthly <- tbl(tidy_finance, "crsp_monthly") |>
  collect()

crsp_monthly <- crsp_monthly |>
  select(
    permno, gvkey, month, ret_excess,
    mktcap, mktcap_lag, exchange
  ) |>
  drop_na()
```

Further, we utilize accounting data. The most common source of accounting data is Compustat. We only need book equity data in this application, which we select from our database. Additionally, we convert the variable `datadate` to its monthly value, as we only consider monthly returns here and do not need to account for the exact date. To achieve this, we use the function `floor_date()`.

```
compustat <- tbl(tidy_finance, "compustat") |>
  collect()

be <- compustat |>
  select(gvkey, datadate, be) |>
  drop_na() |>
  mutate(month = floor_date(ymd(datadate), "month"))
```

9.2 Book-to-Market Ratio

A fundamental problem in handling accounting data is the *look-ahead bias* – we must not include data in forming a portfolio that is not public knowledge at the time. Of course, researchers have more information when looking into the past than agents had at that moment. However, abnormal excess returns from a trading strategy should not rely on an information advantage because the differential cannot be the result of informed agents' trades. Hence, we have to lag accounting information.

We continue to lag market capitalization and firm size by one month. Then, we compute the book-to-market ratio, which relates a firm's book equity to its market equity. Firms with high (low) book-to-market ratio are called value (growth) firms. After matching the accounting and market equity information from the same month, we lag book-to-market by six months. This is a sufficiently conservative approach because accounting information is usually released well before six months pass. However, in the asset pricing literature, even longer lags are used as well.[1]

Having both variables, i.e., firm size lagged by one month and book-to-market lagged by six months, we merge these sorting variables to our returns using the `sorting_date`-column created for this purpose. The final step in our data preparation deals with differences in the frequency of our variables. Returns and firm size are recorded monthly. Yet the accounting information is only released on an annual basis. Hence, we only match book-to-market to one month per year and have eleven empty observations. To solve this frequency issue, we carry the latest book-to-market ratio of each firm to the subsequent months, i.e., we fill the missing observations with the most current report. This is done via the `fill()`-function after sorting by date and firm (which we identify by `permno` and `gvkey`) and on a firm basis (which we do by `group_by()` as usual). We filter out all observations with accounting data that is older than a year. As the last step, we remove all rows with missing entries because the returns cannot be matched to any annual report.

```
me <- crsp_monthly |>
  mutate(sorting_date = month %m+% months(1)) |>
  select(permno, sorting_date, me = mktcap)

bm <- be |>
  inner_join(crsp_monthly |>
    select(month, permno, gvkey, mktcap), by = c("gvkey", "month")) |>
  mutate(
    bm = be / mktcap,
    sorting_date = month %m+% months(6),
    comp_date = sorting_date
  ) |>
  select(permno, gvkey, sorting_date, comp_date, bm)

data_for_sorts <- crsp_monthly |>
  left_join(bm, by = c("permno",
    "gvkey",
    "month" = "sorting_date"
  )) |>
  left_join(me, by = c("permno", "month" = "sorting_date")) |>
  select(
    permno, gvkey, month, ret_excess,
    mktcap_lag, me, bm, exchange, comp_date
  )

data_for_sorts <- data_for_sorts |>
```

[1]The definition of a time lag is another choice a researcher has to make, similar to breakpoint choices as we describe in the Chapter 8 on p-hacking.

```
  arrange(permno, gvkey, month) |>
  group_by(permno, gvkey) |>
  fill(bm, comp_date) |>
  filter(comp_date > month %m-% months(12)) |>
  select(-comp_date) |>
  drop_na()
```

The last step of preparation for the portfolio sorts is the computation of breakpoints. We continue to use the same function allowing for the specification of exchanges to use for the breakpoints. Additionally, we reintroduce the argument `var` into the function for defining different sorting variables via `curly-curly`.

```
assign_portfolio <- function(data, var, n_portfolios, exchanges) {
  breakpoints <- data |>
    filter(exchange %in% exchanges) |>
    summarize(breakpoint = quantile(
      {{ var }},
      probs = seq(0, 1, length.out = n_portfolios + 1),
      na.rm = TRUE
    )) |>
    pull(breakpoint) |>
    as.numeric()

  assigned_portfolios <- data |>
    mutate(portfolio = findInterval({{ var }},
      breakpoints,
      all.inside = TRUE
    )) |>
    pull(portfolio)

  return(assigned_portfolios)
}
```

After these data preparation steps, we present bivariate portfolio sorts on an independent and dependent basis.

9.3 Independent Sorts

Bivariate sorts create portfolios within a two-dimensional space spanned by two sorting variables. It is then possible to assess the return impact of either sorting variable by the return differential from a trading strategy that invests in the portfolios at either end of the respective variables spectrum. We create a five-by-five matrix using book-to-market and firm size as sorting variables in our example below. We end up with 25 portfolios. Since we are interested in the *value premium* (i.e., the return differential between high and low book-to-market firms), we go long the five portfolios of the highest book-to-market firms and short the five portfolios of the lowest book-to-market firms. The five portfolios at each end are due to the size splits we employed alongside the book-to-market splits.

To implement the independent bivariate portfolio sort, we assign monthly portfolios for each of our sorting variables separately to create the variables `portfolio_bm` and `portfolio_me`, respectively. Then, these separate portfolios are combined to the final sort stored in `portfolio_combined`. After assigning the portfolios, we compute the average return within each portfolio for each month. Additionally, we keep the book-to-market portfolio as it makes the computation of the value premium easier. The alternative would be to disaggregate the combined portfolio in a separate step. Notice that we weigh the stocks within each portfolio by their market capitalization, i.e., we decide to value-weight our returns.

```
value_portfolios <- data_for_sorts |>
  group_by(month) |>
  mutate(
    portfolio_bm = assign_portfolio(
      data = cur_data(),
      var = bm,
      n_portfolios = 5,
      exchanges = c("NYSE")
    ),
    portfolio_me = assign_portfolio(
      data = cur_data(),
      var = me,
      n_portfolios = 5,
      exchanges = c("NYSE")
    ),
    portfolio_combined = str_c(portfolio_bm, portfolio_me)
  ) |>
  group_by(month, portfolio_combined) |>
  summarize(
    ret = weighted.mean(ret_excess, mktcap_lag),
    portfolio_bm = unique(portfolio_bm),
    .groups = "drop"
  )
```

Equipped with our monthly portfolio returns, we are ready to compute the value premium. However, we still have to decide how to invest in the five high and the five low book-to-market portfolios. The most common approach is to weigh these portfolios equally, but this is yet another researcher's choice. Then, we compute the return differential between the high and low book-to-market portfolios and show the average value premium.

```
value_premium <- value_portfolios |>
  group_by(month, portfolio_bm) |>
  summarize(ret = mean(ret), .groups = "drop_last") |>
  summarize(value_premium = ret[portfolio_bm == max(portfolio_bm)] -
    ret[portfolio_bm == min(portfolio_bm)])

mean(value_premium$value_premium * 100)
```

```
[1] 0.384
```

The resulting annualized value premium is 4.608 percent.

9.4 Dependent Sorts

In the previous exercise, we assigned the portfolios without considering the second variable in the assignment. This protocol is called independent portfolio sorts. The alternative, i.e., dependent sorts, creates portfolios for the second sorting variable within each bucket of the first sorting variable. In our example below, we sort firms into five size buckets, and within each of those buckets, we assign firms to five book-to-market portfolios. Hence, we have monthly breakpoints that are specific to each size group. The decision between independent and dependent portfolio sorts is another choice for the researcher. Notice that dependent sorts ensure an equal amount of stocks within each portfolio.

To implement the dependent sorts, we first create the size portfolios by calling `assign_portfolio()` with `var = me`. Then, we group our data again by month and by the size portfolio before assigning the book-to-market portfolio. The rest of the implementation is the same as before. Finally, we compute the value premium.

```
value_portfolios <- data_for_sorts |>
  group_by(month) |>
  mutate(portfolio_me = assign_portfolio(
    data = cur_data(),
    var = me,
    n_portfolios = 5,
    exchanges = c("NYSE")
  )) |>
```

```
  group_by(month, portfolio_me) |>
  mutate(
    portfolio_bm = assign_portfolio(
      data = cur_data(),
      var = bm,
      n_portfolios = 5,
      exchanges = c("NYSE")
    ),
    portfolio_combined = str_c(portfolio_bm, portfolio_me)
  ) |>
  group_by(month, portfolio_combined) |>
  summarize(
    ret = weighted.mean(ret_excess, mktcap_lag),
    portfolio_bm = unique(portfolio_bm),
    .groups = "drop"
  )

value_premium <- value_portfolios |>
  group_by(month, portfolio_bm) |>
  summarize(ret = mean(ret), .groups = "drop_last") |>
  summarize(value_premium = ret[portfolio_bm == max(portfolio_bm)] -
    ret[portfolio_bm == min(portfolio_bm)])

mean(value_premium$value_premium * 100)
```

```
[1] 0.329
```

The value premium from dependent sorts is 3.948 percent per year.

Overall, we show how to conduct bivariate portfolio sorts in this chapter. In one case, we sort the portfolios independently of each other. Yet we also discuss how to create dependent portfolio sorts. Along the lines of Chapter 8, we see how many choices a researcher has to make to implement portfolio sorts, and bivariate sorts increase the number of choices.

9.5 Exercises

1. In Chapter 8, we examine the distribution of market equity. Repeat this analysis for book equity and the book-to-market ratio (alongside a plot of the breakpoints, i.e., deciles).

2. When we investigate the portfolios, we focus on the returns exclusively. However, it is also of interest to understand the characteristics of the portfolios. Write a function to compute the average characteristics for size and book-to-market across the 25 independently and dependently sorted portfolios.
3. As for the size premium, also the value premium constructed here does not follow Fama and French (1993). Implement a p-hacking setup as in Chapter 8 to find a premium that comes closest to their HML premium.

10

Replicating Fama and French Factors

In this chapter, we provide a replication of the famous Fama and French factor portfolios. The Fama and French three-factor model (see Fama and French, 1993) is a cornerstone of asset pricing. On top of the market factor represented by the traditional CAPM beta, the model includes the size and value factors to explain the cross section of returns. We introduce both factors in Chapter 9, and their definition remains the same. Size is the SMB factor (small-minus-big) that is long small firms and short large firms. The value factor is HML (high-minus-low) and is long in high book-to-market firms and short in low book-to-market counterparts.

The current chapter relies on this set of packages.

```
library(tidyverse)
library(RSQLite)
library(lubridate)
```

10.1 Data Preparation

We use CRSP and Compustat as data sources, as we need exactly the same variables to compute the size and value factors in the way Fama and French do it. Hence, there is nothing new below and we only load data from our `SQLite`-database introduced in Chapters 2–4.

```
tidy_finance <- dbConnect(
  SQLite(),
  "data/tidy_finance.sqlite",
  extended_types = TRUE
)

crsp_monthly <- tbl(tidy_finance, "crsp_monthly") |>
  collect()

data_ff <- crsp_monthly |>
```

DOI: 10.1201/b23237-10

```
  select(
    permno, gvkey, month, ret_excess,
    mktcap, mktcap_lag, exchange
  ) |>
  drop_na()

compustat <- tbl(tidy_finance, "compustat") |>
  collect()

be <- compustat |>
  select(gvkey, datadate, be) |>
  drop_na()

factors_ff_monthly <- tbl(tidy_finance, "factors_ff_monthly") |>
  collect()
```

Yet when we start merging our data set for computing the premiums, there are a few differences to Chapter 9. First, Fama and French form their portfolios in June of year t, whereby the returns of July are the first monthly return for the respective portfolio. For firm size, they consequently use the market capitalization recorded for June. It is then held constant until June of year $t+1$.

Second, Fama and French also have a different protocol for computing the book-to-market ratio. They use market equity as of the end of year $t-1$ and the book equity reported in year $t-1$, i.e., the `datadate` is within the last year. Hence, the book-to-market ratio can be based on accounting information that is up to 18 months old. Market equity also does not necessarily reflect the same time point as book equity.

To implement all these time lags, we again employ the temporary `sorting_date`-column. Notice that when we combine the information, we want to have a single observation per year and stock since we are only interested in computing the breakpoints held constant for the entire year. We ensure this by a call of `distinct()` at the end of the chunk below.

```
me_ff <- data_ff |>
  filter(month(month) == 6) |>
  mutate(sorting_date = month %m+% months(1)) |>
  select(permno, sorting_date, me_ff = mktcap)

me_ff_dec <- data_ff |>
  filter(month(month) == 12) |>
  mutate(sorting_date = ymd(str_c(year(month) + 1, "0701)"))) |>
  select(permno, gvkey, sorting_date, bm_me = mktcap)

bm_ff <- be |>
  mutate(sorting_date = ymd(str_c(year(datadate) + 1, "0701"))) |>
```

```
  select(gvkey, sorting_date, bm_be = be) |>
  drop_na() |>
  inner_join(me_ff_dec, by = c("gvkey", "sorting_date")) |>
  mutate(bm_ff = bm_be / bm_me) |>
  select(permno, sorting_date, bm_ff)

variables_ff <- me_ff |>
  inner_join(bm_ff, by = c("permno", "sorting_date")) |>
  drop_na() |>
  distinct(permno, sorting_date, .keep_all = TRUE)
```

10.2 Portfolio Sorts

Next, we construct our portfolios with an adjusted `assign_portfolio()` function. Fama and French rely on NYSE-specific breakpoints, they form two portfolios in the size dimension at the median and three portfolios in the dimension of book-to-market at the 30%- and 70%-percentiles, and they use independent sorts. The sorts for book-to-market require an adjustment to the function in Chapter 9 because the `seq()` we would produce does not produce the right breakpoints. Instead of `n_portfolios`, we now specify `percentiles`, which take the breakpoint-sequence as an object specified in the function's call. Specifically, we give `percentiles = c(0, 0.3, 0.7, 1)` to the function. Additionally, we perform an `inner_join()` with our return data to ensure that we only use traded stocks when computing the breakpoints as a first step.

```
assign_portfolio <- function(data, var, percentiles) {
  breakpoints <- data |>
    filter(exchange == "NYSE") |>
    summarize(breakpoint = quantile(
      {{ var }},
      probs = {{ percentiles }},
      na.rm = TRUE
    )) |>
    pull(breakpoint) |>
    as.numeric()

  assigned_portfolios <- data |>
    mutate(portfolio = findInterval({{ var }},
      breakpoints,
      all.inside = TRUE
    )) |>
```

```
    pull(portfolio)

  return(assigned_portfolios)
}

portfolios_ff <- variables_ff |>
  inner_join(data_ff, by = c("permno" = "permno", "sorting_date" = "month")) |>
  group_by(sorting_date) |>
  mutate(
    portfolio_me = assign_portfolio(
      data = cur_data(),
      var = me_ff,
      percentiles = c(0, 0.5, 1)
    ),
    portfolio_bm = assign_portfolio(
      data = cur_data(),
      var = bm_ff,
      percentiles = c(0, 0.3, 0.7, 1)
    )
  ) |>
  select(permno, sorting_date, portfolio_me, portfolio_bm)
```

Next, we merge the portfolios to the return data for the rest of the year. To implement this step, we create a new column `sorting_date` in our return data by setting the date to sort on to July of $t-1$ if the month is June (of year t) or earlier or to July of year t if the month is July or later.

```
portfolios_ff <- data_ff |>
  mutate(sorting_date = case_when(
    month(month) <= 6 ~ ymd(str_c(year(month) - 1, "0701")),
    month(month) >= 7 ~ ymd(str_c(year(month), "0701"))
  )) |>
  inner_join(portfolios_ff, by = c("permno", "sorting_date"))
```

10.3 Fama and French Factor Returns

Equipped with the return data and the assigned portfolios, we can now compute the value-weighted average return for each of the six portfolios. Then, we form the Fama and French factors. For the size factor (i.e., SMB), we go long in the three small portfolios and short the three large portfolios by taking an average across either group.

For the value factor (i.e., HML), we go long in the two high book-to-market portfolios and short the two low book-to-market portfolios, again weighting them equally.

```
factors_ff_monthly_replicated <- portfolios_ff |>
  mutate(portfolio = str_c(portfolio_me, portfolio_bm)) |>
  group_by(portfolio, month) |>
  summarize(
    ret = weighted.mean(ret_excess, mktcap_lag), .groups = "drop",
    portfolio_me = unique(portfolio_me),
    portfolio_bm = unique(portfolio_bm)
  ) |>
  group_by(month) |>
  summarize(
    smb_replicated = mean(ret[portfolio_me == 1]) -
      mean(ret[portfolio_me == 2]),
    hml_replicated = mean(ret[portfolio_bm == 3]) -
      mean(ret[portfolio_bm == 1])
  )
```

10.4 Replication Evaluation

In the previous section, we replicated the size and value premiums following the procedure outlined by Fama and French. However, we did not follow their procedure strictly. The final question is then: how close did we get? We answer this question by looking at the two time-series estimates in a regression analysis using `lm()`. If we did a good job, then we should see a non-significant intercept (rejecting the notion of systematic error), a coefficient close to 1 (indicating a high correlation), and an adjusted R-squared close to 1 (indicating a high proportion of explained variance).

```
test <- factors_ff_monthly |>
  inner_join(factors_ff_monthly_replicated, by = "month") |>
  mutate(
    smb_replicated = round(smb_replicated, 4),
    hml_replicated = round(hml_replicated, 4)
  )
```

The results for the SMB factor are quite convincing as all three criteria outlined above are met and the coefficient and R-squared are at 99%.

```
summary(lm(smb ~ smb_replicated, data = test))
```

```
Call:
lm(formula = smb ~ smb_replicated, data = test)

Residuals:
      Min        1Q    Median        3Q       Max
-0.020352 -0.001590  0.000007  0.001563  0.014636

Coefficients:
                Estimate Std. Error t value Pr(>|t|)
(Intercept)    -0.000129   0.000131   -0.98     0.33
smb_replicated  0.995395   0.004341  229.31   <2e-16 ***
---
Signif. codes:  0 '***' 0.001 '**' 0.01 '*' 0.05 '.' 0.1 ' ' 1

Residual standard error: 0.00352 on 724 degrees of freedom
Multiple R-squared:  0.986, Adjusted R-squared:  0.986
F-statistic: 5.26e+04 on 1 and 724 DF,  p-value: <2e-16
```

The replication of the HML factor is also a success, although at a slightly lower level with coefficient and R-squared around 95%.

```
summary(lm(hml ~ hml_replicated, data = test))
```

```
Call:
lm(formula = hml ~ hml_replicated, data = test)

Residuals:
      Min        1Q    Median        3Q       Max
-0.023350 -0.002984 -0.000091  0.002280  0.028783

Coefficients:
               Estimate Std. Error t value Pr(>|t|)
(Intercept)    0.000318   0.000219    1.45     0.15
hml_replicated 0.965370   0.007478  129.09   <2e-16 ***
---
Signif. codes:  0 '***' 0.001 '**' 0.01 '*' 0.05 '.' 0.1 ' ' 1

Residual standard error: 0.00587 on 724 degrees of freedom
Multiple R-squared:  0.958, Adjusted R-squared:  0.958
F-statistic: 1.67e+04 on 1 and 724 DF,  p-value: <2e-16
```

The evidence hence allows us to conclude that we did a relatively good job in replicating the original Fama-French premiums, although we cannot see their underlying code.

From our perspective, a perfect match is only possible with additional information from the maintainers of the original data.

10.5 Exercises

1. Fama and French (1993) claim that their sample excludes firms until they have appeared in Compustat for two years. Implement this additional filter and compare the improvements of your replication effort.
2. On his homepage, Kenneth French[1] provides instructions on how to construct the most common variables used for portfolio sorts. Pick one of them, e.g. OP (operating profitability) and try to replicate the portfolio sort return time series provided on his homepage.

[1]https://mba.tuck.dartmouth.edu/pages/faculty/ken.french/Data_Library/variable_definitions.html

11

Fama-MacBeth Regressions

In this chapter, we present a simple implementation of Fama and MacBeth (1973), a regression approach commonly called Fama-MacBeth regressions. Fama-MacBeth regressions are widely used in empirical asset pricing studies. We use individual stocks as test assets to estimate the risk premium associated with the three factors included in Fama and French (1993).

Researchers use the two-stage regression approach to estimate risk premiums in various markets, but predominately in the stock market. Essentially, the two-step Fama-MacBeth regressions exploit a linear relationship between expected returns and exposure to (priced) risk factors. The basic idea of the regression approach is to project asset returns on factor exposures or characteristics that resemble exposure to a risk factor in the cross-section in each time period. Then, in the second step, the estimates are aggregated across time to test if a risk factor is priced. In principle, Fama-MacBeth regressions can be used in the same way as portfolio sorts introduced in previous chapters.

The Fama-MacBeth procedure is a simple two-step approach: The first step uses the exposures (characteristics) as explanatory variables in T cross-sectional regressions. For example, if $r_{i,t+1}$ denote the excess returns of asset i in month $t+1$, then the famous Fama-French three factor model implies the following return generating process (see also Campbell et al., 1998):

$$r_{i,t+1} = \alpha_i + \lambda^M_t \beta^M_{i,t} + \lambda^{SMB}_t \beta^{SMB}_{i,t} + \lambda^{HML}_t \beta^{HML}_{i,t} + \epsilon_{i,t}.$$

Here, we are interested in the compensation λ^f_t for the exposure to each risk factor $\beta^f_{i,t}$ at each time point, i.e., the risk premium. Note the terminology: $\beta^f_{i,t}$ is a asset-specific characteristic, e.g., a factor exposure or an accounting variable. *If* there is a linear relationship between expected returns and the characteristic in a given month, we expect the regression coefficient to reflect the relationship, i.e., $\lambda^f_t \neq 0$.

In the second step, the time-series average $\frac{1}{T}\sum_{t=1}^{T}\hat{\lambda}^f_t$ of the estimates $\hat{\lambda}^f_t$ can then be interpreted as the risk premium for the specific risk factor f. We follow Zaffaroni and Zhou (2022) and consider the standard cross-sectional regression to predict future returns. If the characteristics are replaced with time $t+1$ variables, then the regression approach captures risk attributes rather than risk premiums.

Before we move to the implementation, we want to highlight that the characteristics, e.g., $\hat{\beta}^f_i$, are often estimated in a separate step before applying the actual Fama-MacBeth methodology. You can think of this as a *step 0*. You might thus worry that

DOI: 10.1201/b23237-11

the errors of $\hat{\beta}_i^f$ impact the risk premiums' standard errors. Measurement error in $\hat{\beta}_i^f$ indeed affects the risk premium estimates, i.e., they lead to biased estimates. The literature provides adjustments for this bias (see, e.g., Shanken, 1992; Kim, 1995; Chen et al., 2015, among others) but also shows that the bias goes to zero as $T \to \infty$. We refer to Gagliardini et al. (2016) for an in-depth discussion also covering the case of time-varying betas. Moreover, if you plan to use Fama-MacBeth regressions with individual stocks: Hou et al. (2020) advocates using weighed-least squares to estimate the coefficients such that they are not biased toward small firms. Without this adjustment, the high number of small firms would drive the coefficient estimates.

The current chapter relies on this set of packages.

```
library(tidyverse)
library(RSQLite)
library(lubridate)
library(sandwich)
library(broom)
```

Compared to previous chapters, we introduce the `broom` package (Robinson et al., 2022) to tidy the estimation output of many estimated linear models.

11.1 Data Preparation

We illustrate Fama and MacBeth (1973) with the monthly CRSP sample and use three characteristics to explain the cross-section of returns: market capitalization, the book-to-market ratio, and the CAPM beta (i.e., the covariance of the excess stock returns with the market excess returns). We collect the data from our `SQLite`-database introduced in Chapters 2–4.

```
tidy_finance <- dbConnect(
  SQLite(),
  "data/tidy_finance.sqlite",
  extended_types = TRUE
)

crsp_monthly <- tbl(tidy_finance, "crsp_monthly") |>
  collect()

compustat <- tbl(tidy_finance, "compustat") |>
  collect()
```

```
beta <- tbl(tidy_finance, "beta") |>
  collect()
```

We use the Compustat and CRSP data to compute the book-to-market ratio and the (log) market capitalization. Furthermore, we also use the CAPM betas based on monthly returns we computed in the previous chapters.

```
characteristics <- compustat |>
  mutate(month = floor_date(ymd(datadate), "month")) |>
  left_join(crsp_monthly, by = c("gvkey", "month")) |>
  left_join(beta, by = c("permno", "month")) |>
  transmute(gvkey,
    bm = be / mktcap,
    log_mktcap = log(mktcap),
    beta = beta_monthly,
    sorting_date = month %m+% months(6)
  )

data_fama_macbeth <- crsp_monthly |>
  left_join(characteristics, by = c("gvkey", "month" = "sorting_date")) |>
  group_by(permno) |>
  arrange(month) |>
  fill(c(beta, bm, log_mktcap), .direction = "down") |>
  ungroup() |>
  left_join(crsp_monthly |>
    select(permno, month, ret_excess_lead = ret) |>
    mutate(month = month %m-% months(1)),
  by = c("permno", "month")
  ) |>
  select(permno, month, ret_excess_lead, beta, log_mktcap, bm) |>
  drop_na()
```

11.2 Cross-sectional Regression

Next, we run the cross-sectional regressions with the characteristics as explanatory variables for each month.
We regress the returns of the test assets at a particular time point on the characteristisc of each asset. By doing so, we get an estimate of the risk premiums $\hat{\lambda}_t^f$ for each point in time.

```
risk_premiums <- data_fama_macbeth |>
  nest(data = c(ret_excess_lead, beta, log_mktcap, bm, permno)) |>
  mutate(estimates = map(
    data,
    ~ tidy(lm(ret_excess_lead ~ beta + log_mktcap + bm, data = .x))
  )) |>
  unnest(estimates)
```

11.3 Time-Series Aggregation

Now that we have the risk premiums' estimates for each period, we can average across the time-series dimension to get the expected risk premium for each characteristic. Similarly, we manually create the t-test statistics for each regressor, which we can then compare to usual critical values of 1.96 or 2.576 for two-tailed significance tests.

```
price_of_risk <- risk_premiums |>
  group_by(factor = term) |>
  summarize(
    risk_premium = mean(estimate) * 100,
    t_statistic = mean(estimate) / sd(estimate) * sqrt(n())
  )
```

It is common to adjust for autocorrelation when reporting standard errors of risk premiums. As in Chapter 7, the typical procedure for this is computing Newey and West (1987) standard errors. We again recommend the data-driven approach of Newey and West (1994) using the `NeweyWest()` function, but note that you can enforce the typical 6 lag settings via `NeweyWest(., lag = 6, prewhite = FALSE)`.

```
regressions_for_newey_west <- risk_premiums |>
  select(month, factor = term, estimate) |>
  nest(data = c(month, estimate)) |>
  mutate(
    model = map(data, ~ lm(estimate ~ 1, .)),
    mean = map(model, tidy)
  )

price_of_risk_newey_west <- regressions_for_newey_west |>
  mutate(newey_west_se = map_dbl(model, ~ sqrt(NeweyWest(.)))) |>
  unnest(mean) |>
```

```
  mutate(t_statistic_newey_west = estimate / newey_west_se) |>
  select(factor,
    risk_premium = estimate,
    t_statistic_newey_west
  )

left_join(price_of_risk,
  price_of_risk_newey_west |>
    select(factor, t_statistic_newey_west),
  by = "factor"
)
```

```
# A tibble: 4 x 4
  factor      risk_premium t_statistic t_statistic_newey_west
  <chr>              <dbl>       <dbl>                  <dbl>
1 (Intercept)      1.70         6.68                   5.76
2 beta             0.00763      0.0730                 0.0651
3 bm               0.141        3.03                   2.59
4 log_mktcap      -0.114       -3.20                  -2.94
```

Finally, let us interpret the results. Stocks with higher book-to-market ratios earn higher expected future returns, which is in line with the value premium. The negative value for log market capitalization reflects the size premium for smaller stocks. Consistent with results from earlier chapters, we detect no relation between beta and future stock returns.

11.4 Exercises

1. Download a sample of test assets from Kenneth French's homepage and reevaluate the risk premiums for industry portfolios instead of individual stocks.
2. Use individual stocks with weighted-least squares based on a firm's size as suggested by Hou et al. (2020). Then, repeat the Fama-MacBeth regressions without the weighting scheme adjustment but drop the smallest 20 percent of firms each month. Compare the results of the three approaches.
3. Implement a rolling-window regression for the time-series estimation of the factor exposure. Skip one month after each rolling period before including the exposures in the cross-sectional regression to avoid a look-ahead bias. Then, adapt the cross-sectional regression and compute the average risk premiums.

Part IV

Modeling & Machine Learning

12

Fixed Effects and Clustered Standard Errors

In this chapter, we provide an intuitive introduction to the two popular concepts of *fixed effects regressions* and *clustered standard errors*. When working with regressions in empirical finance, you will sooner or later be confronted with discussions around how you deal with omitted variables bias and dependence in your residuals. The concepts we introduce in this chapter are designed to address such concerns.

We focus on a classical panel regression common to the corporate finance literature (e.g., Fazzari et al., 1988; Erickson and Whited, 2012; Gulen and Ion, 2015): firm investment modeled as a function that increases in firm cash flow and firm investment opportunities.

Typically, this investment regression uses quarterly balance sheet data provided via Compustat because it allows for richer dynamics in the regressors and more opportunities to construct variables. As we focus on the implementation of fixed effects and clustered standard errors, we use the annual Compustat data from our previous chapters and leave the estimation using quarterly data as an exercise. We demonstrate below that the regression based on annual data yields qualitatively similar results to estimations based on quarterly data from the literature, namely confirming the positive relationships between investment and the two regressors.

The current chapter relies on the following set of packages.

```
library(tidyverse)
library(RSQLite)
library(lubridate)
library(fixest)
```

Compared to previous chapters, we introduce `fixest` (Bergé, 2018) for the fixed effects regressions, the implementation of standard error clusters, and tidy estimation output.

DOI: 10.1201/b23237-12

12.1 Data Preparation

We use CRSP and annual Compustat as data sources from our `SQLite`-database introduced in Chapters 2–4. In particular, Compustat provides balance sheet and income statement data on a firm level, while CRSP provides market valuations.

```
tidy_finance <- dbConnect(
  SQLite(),
  "data/tidy_finance.sqlite",
  extended_types = TRUE
)

crsp <- tbl(tidy_finance, "crsp_monthly") |>
  collect()

compustat <- tbl(tidy_finance, "compustat") |>
  collect()
```

The classical investment regressions model the capital investment of a firm as a function of operating cash flows and Tobin's q, a measure of a firm's investment opportunities. We start by constructing investment and cash flows which are usually normalized by lagged total assets of a firm. In the following code chunk, we construct a *panel* of firm-year observations, so we have both cross-sectional information on firms as well as time-series information for each firm.

```
data_investment <- compustat |>
  mutate(month = floor_date(datadate, "month")) |>
  left_join(compustat |>
    select(gvkey, year, at_lag = at) |>
    mutate(year = year + 1),
  by = c("gvkey", "year")
  ) |>
  filter(at > 0, at_lag > 0) |>
  mutate(
    investment = capx / at_lag,
    cash_flows = oancf / at_lag
  )

data_investment <- data_investment |>
  left_join(data_investment |>
    select(gvkey, year, investment_lead = investment) |>
    mutate(year = year - 1),
```

```
  by = c("gvkey", "year")
  )
```

Tobin's q is the ratio of the market value of capital to its replacement costs. It is one of the most common regressors in corporate finance applications (e.g., Fazzari et al., 1988; Erickson and Whited, 2012). We follow the implementation of Gulen and Ion (2015) and compute Tobin's q as the market value of equity (`mktcap`) plus the book value of assets (`at`) minus book value of equity (`be`) plus deferred taxes (`txdb`), all divided by book value of assets (`at`). Finally, we only keep observations where all variables of interest are non-missing, and the reported book value of assets is strictly positive.

```
data_investment <- data_investment |>
  left_join(crsp |>
    select(gvkey, month, mktcap),
  by = c("gvkey", "month")
  ) |>
  mutate(tobins_q = (mktcap + at - be + txdb) / at)

data_investment <- data_investment |>
  select(gvkey, year, investment_lead, cash_flows, tobins_q) |>
  drop_na()
```

As the variable construction typically leads to extreme values that are most likely related to data issues (e.g., reporting errors), many papers include winsorization of the variables of interest. Winsorization involves replacing values of extreme outliers with quantiles on the respective end. The following function implements the winsorization for any percentage cut that should be applied on either end of the distributions. In the specific example, we winsorize the main variables (`investment`, `cash_flows`, and `tobins_q`) at the 1 percent level.

```
winsorize <- function(x, cut) {
  x <- replace(
    x,
    x > quantile(x, 1 - cut, na.rm = T),
    quantile(x, 1 - cut, na.rm = T)
  )
  x <- replace(
    x,
    x < quantile(x, cut, na.rm = T),
    quantile(x, cut, na.rm = T)
  )
  return(x)
}
```

```
data_investment <- data_investment |>
  mutate(across(
    c(investment_lead, cash_flows, tobins_q),
    ~ winsorize(., 0.01)
  ))
```

Before proceeding to any estimations, we highly recommend tabulating summary statistics of the variables that enter the regression. These simple tables allow you to check the plausibility of your numerical variables, as well as spot any obvious errors or outliers. Additionally, for panel data, plotting the time series of the variable's mean and the number of observations is a useful exercise to spot potential problems.

```
data_investment |>
  pivot_longer(
    cols = c(investment_lead, cash_flows, tobins_q),
    names_to = "measure"
  ) |>
  group_by(measure) |>
  summarize(
    mean = mean(value),
    sd = sd(value),
    min = min(value),
    q05 = quantile(value, 0.05),
    q50 = quantile(value, 0.50),
    q95 = quantile(value, 0.95),
    max = max(value),
    n = n(),
    .groups = "drop"
  )
```

```
# A tibble: 3 x 9
  measure       mean     sd    min       q05    q50   q95    max      n
  <chr>        <dbl>  <dbl>  <dbl>     <dbl>  <dbl> <dbl>  <dbl>  <int>
1 cash_flows  0.0145 0.266  -1.50  -4.57e-1 0.0649 0.273  0.480 124179
2 investmen~  0.0584 0.0778  0      7.27e-4 0.0333 0.208  0.467 124179
3 tobins_q    1.99   1.69    0.571  7.92e-1 1.38   5.33  10.9   124179
```

12.2 Fixed Effects

To illustrate fixed effects regressions, we use the `fixest` package, which is both computationally powerful and flexible with respect to model specifications. We start out with the basic investment regression using the simple model

$$\text{Investment}_{i,t+1} = \alpha + \beta_1 \text{Cash Flows}_{i,t} + \beta_2 \text{Tobin's q}_{i,t} + \varepsilon_{i,t},$$

where ε_t is i.i.d. normally distributed across time and firms. We use the `feols()`-function to estimate the simple model so that the output has the same structure as the other regressions below, but you could also use `lm()`.

```
model_ols <- feols(
  fml = investment_lead ~ cash_flows + tobins_q,
  se = "iid",
  data = data_investment
)
model_ols
```

```
OLS estimation, Dep. Var.: investment_lead
Observations: 124,179
Standard-errors: IID
            Estimate Std. Error t value  Pr(>|t|)
(Intercept)  0.04243   0.000342   124.1 < 2.2e-16 ***
cash_flows   0.05143   0.000835    61.6 < 2.2e-16 ***
tobins_q     0.00767   0.000132    58.2 < 2.2e-16 ***
---
Signif. codes:  0 '***' 0.001 '**' 0.01 '*' 0.05 '.' 0.1 ' ' 1
RMSE: 0.076041   Adj. R2: 0.044436
```

As expected, the regression output shows significant coefficients for both variables. Higher cash flows and investment opportunities are associated with higher investment. However, the simple model actually may have a lot of omitted variables, so our coefficients are most likely biased. As there is a lot of unexplained variation in our simple model (indicated by the rather low adjusted R-squared), the bias in our coefficients is potentially severe, and the true values could be above or below zero. Note that there are no clear cutoffs to decide when an R-squared is high or low, but it depends on the context of your application and on the comparison of different models for the same data.

One way to tackle the issue of omitted variable bias is to get rid of as much unexplained variation as possible by including *fixed effects* – i.e., model parameters that are fixed for specific groups (e.g., Wooldridge, 2010). In essence, each group has its own mean in fixed effects regressions. The simplest group that we can form in the investment regression is the firm level. The firm fixed effects regression is then

$$\text{Investment}_{i,t+1} = \alpha_i + \beta_1 \text{Cash Flows}_{i,t} + \beta_2 \text{Tobin's q}_{i,t} + \varepsilon_{i,t},$$

where α_i is the firm fixed effect and captures the firm-specific mean investment across all years. In fact, you could also compute firms' investments as deviations from the firms' average investments and estimate the model without the fixed effects. The idea of the firm fixed effect is to remove the firm's average investment, which might be affected by firm-specific variables that you do not observe. For example, firms in a specific industry might invest more on average. Or you observe a young firm with large investments but only small concurrent cash flows, which will only happen in a few years. This sort of variation is unwanted because it is related to unobserved variables that can bias your estimates in any direction.

To include the firm fixed effect, we use `gvkey` (Compustat's firm identifier) as follows:

```
model_fe_firm <- feols(
  investment_lead ~ cash_flows + tobins_q | gvkey,
  se = "iid",
  data = data_investment
)
model_fe_firm
```

```
OLS estimation, Dep. Var.: investment_lead
Observations: 124,179
Fixed-effects: gvkey: 13,899
Standard-errors: IID
           Estimate Std. Error t value  Pr(>|t|)
cash_flows   0.0146   0.000963    15.1 < 2.2e-16 ***
tobins_q     0.0113   0.000136    82.6 < 2.2e-16 ***
---
Signif. codes:  0 '***' 0.001 '**' 0.01 '*' 0.05 '.' 0.1 ' ' 1
RMSE: 0.05008     Adj. R2: 0.533298
                Within R2: 0.059424
```

The regression output shows a lot of unexplained variation at the firm level that is taken care of by including the firm fixed effect as the adjusted R-squared rises above 50%. In fact, it is more interesting to look at the within R-squared that shows the explanatory power of a firm's cash flow and Tobin's q *on top* of the average investment of each firm. We can also see that the coefficients changed slightly in magnitude but not in sign.

There is another source of variation that we can get rid of in our setting: average investment across firms might vary over time due to macroeconomic factors that affect all firms, such as economic crises. By including year fixed effects, we can take out the effect of unobservables that vary over time. The two-way fixed effects regression is then

$$\text{Investment}_{i,t+1} = \alpha_i + \alpha_t + \beta_1 \text{Cash Flows}_{i,t} + \beta_2 \text{Tobin's q}_{i,t} + \varepsilon_{i,t},$$

where α_t is the time fixed effect. Here you can think of higher investments during an economic expansion with simultaneously high cash flows.

```
model_fe_firmyear <- feols(
  investment_lead ~ cash_flows + tobins_q | gvkey + year,
  se = "iid",
  data = data_investment
)
model_fe_firmyear
```

```
OLS estimation, Dep. Var.: investment_lead
Observations: 124,179
Fixed-effects: gvkey: 13,899,  year: 34
Standard-errors: IID
           Estimate Std. Error t value  Pr(>|t|)
cash_flows   0.0182   0.000941    19.3 < 2.2e-16 ***
tobins_q     0.0102   0.000135    75.5 < 2.2e-16 ***
---
Signif. codes:  0 '***' 0.001 '**' 0.01 '*' 0.05 '.' 0.1 ' ' 1
RMSE: 0.048805     Adj. R2: 0.556615
                 Within R2: 0.051545
```

The inclusion of time fixed effects did only marginally affect the R-squared and the coefficients, which we can interpret as a good thing as it indicates that the coefficients are not driven by an omitted variable that varies over time.

How can we further improve the robustness of our regression results? Ideally, we want to get rid of unexplained variation at the firm-year level, which means we need to include more variables that vary across firm *and* time and are likely correlated with investment. Note that we cannot include firm-year fixed effects in our setting because then cash flows and Tobin's q are colinear with the fixed effects, and the estimation becomes void.

Before we discuss the properties of our estimation errors, we want to point out that regression tables are at the heart of every empirical analysis, where you compare multiple models. Fortunately, the `etable()` provides a convenient way to tabulate the regression output (with many parameters to customize and even print the output in LaTeX). We recommend printing t-statistics rather than standard errors in regression tables because the latter are typically very hard to interpret across coefficients that vary in size. We also do not print p-values because they are sometimes misinterpreted to signal the importance of observed effects (Wasserstein and Lazar, 2016). The t-statistics provide a consistent way to interpret changes in estimation uncertainty across different model specifications.

```
etable(model_ols, model_fe_firm, model_fe_firmyear,
  coefstat = "tstat"
)
```

```
                       model_ols     model_fe_firm
Dependent Var.:   investment_lead  investment_lead

(Intercept)      0.0424*** (124.1)
cash_flows       0.0514*** (61.56) 0.0146*** (15.14)
tobins_q         0.0077*** (58.17) 0.0113*** (82.60)
Fixed-Effects:   ----------------- -----------------
gvkey                           No               Yes
year                            No                No
_______________  _________________ _________________
VCOV type                      IID               IID
Observations               124,179           124,179
R2                         0.04445           0.58554
Within R2                       --           0.05942

                 model_fe_firmyear
Dependent Var.:    investment_lead

(Intercept)
cash_flows       0.0182*** (19.30)
tobins_q         0.0102*** (75.51)
Fixed-Effects:   -----------------
gvkey                          Yes
year                           Yes
_______________  _________________
VCOV type                      IID
Observations               124,179
R2                         0.60636
Within R2                  0.05155
---
Signif. codes: 0 '***' 0.001 '**' 0.01 '*' 0.05 '.' 0.1 ' ' 1
```

12.3 Clustering Standard Errors

Apart from biased estimators, we usually have to deal with potentially complex dependencies of our residuals with each other. Such dependencies in the residuals invalidate the i.i.d. assumption of OLS and lead to biased standard errors. With biased OLS standard errors, we cannot reliably interpret the statistical significance of our estimated coefficients.

In our setting, the residuals may be correlated across years for a given firm (time-series dependence), or, alternatively, the residuals may be correlated across different firms (cross-section dependence). One of the most common approaches to dealing with such dependence is the use of *clustered standard errors* (Petersen, 2008). The idea behind clustering is that the correlation of residuals *within* a cluster can be of any form. As the number of clusters grows, the cluster-robust standard errors become consistent (Donald and Lang, 2007; Wooldridge, 2010). A natural requirement for clustering standard errors in practice is hence a sufficiently large number of clusters. Typically, around at least 30 to 50 clusters are seen as sufficient (Cameron et al., 2011).

Instead of relying on the iid assumption, we can use the cluster option in the `feols`-function as above. The code chunk below applies both one-way clustering by firm as well as two-way clustering by firm and year.

```
model_cluster_firm <- feols(
  investment_lead ~ cash_flows + tobins_q | gvkey + year,
  cluster = "gvkey",
  data = data_investment
)

model_cluster_firmyear <- feols(
  investment_lead ~ cash_flows + tobins_q | gvkey + year,
  cluster = c("gvkey", "year"),
  data = data_investment
)
```

The table below shows the comparison of the different assumptions behind the standard errors. In the first column, we can see highly significant coefficients on both cash flows and Tobin's q. By clustering the standard errors on the firm level, the t-statistics of both coefficients drop in half, indicating a high correlation of residuals within firms. If we additionally cluster by year, we see a drop, particularly for Tobin's q, again. Even after relaxing the assumptions behind our standard errors, both coefficients are still comfortably significant as the t statistics are well above the usual critical values of 1.96 or 2.576 for two-tailed significance tests.

```
etable(model_fe_firmyear, model_cluster_firm, model_cluster_firmyear,
  coefstat = "tstat"
)
```

```
                 model_fe_firmyear model_cluster_f..
Dependent Var.:    investment_lead   investment_lead


cash_flows       0.0182*** (19.30) 0.0182*** (11.18)
tobins_q         0.0102*** (75.51) 0.0102*** (35.63)
Fixed-Effects:   ----------------- -----------------
gvkey                          Yes               Yes
year                           Yes               Yes
```

```
--------------- ----------------- -----------------
VCOV type                     IID          by: gvkey
Observations              124,179            124,179
R2                        0.60636            0.60636
Within R2                 0.05155            0.05155

                 model_cluster_f..
Dependent Var.:    investment_lead

cash_flows       0.0182*** (9.459)
tobins_q         0.0102*** (16.36)
Fixed-Effects:   -----------------
gvkey                          Yes
year                           Yes
--------------- -----------------
VCOV type          by: gvkey & year
Observations              124,179
R2                        0.60636
Within R2                 0.05155
---
Signif. codes: 0 '***' 0.001 '**' 0.01 '*' 0.05 '.' 0.1 ' ' 1
```

Inspired by Abadie et al. (2017), we want to close this chapter by highlighting that choosing the right dimensions for clustering is a design problem. Even if the data is informative about whether clustering matters for standard errors, they do not tell you whether you should adjust the standard errors for clustering. Clustering at too aggregate levels can hence lead to unnecessarily inflated standard errors.

12.4 Exercises

1. Estimate the two-way fixed effects model with two-way clustered standard errors using quarterly Compustat data from WRDS. Note that you can access quarterly data via `tbl(wrds, in_schema("comp", "fundq"))`.
2. Following Peters and Taylor (2017), compute Tobin's q as the market value of outstanding equity `mktcap` plus the book value of debt (`dltt` + `dlc`) minus the current assets `atc` and everything divided by the book value of property, plant and equipment `ppegt`. What is the correlation between the measures of Tobin's q? What is the impact on the two-way fixed effects regressions?

13

Difference in Differences

In this chapter, we illustrate the concept of *difference in differences* (DD) estimators by evaluating the effects of climate change regulation on the pricing of bonds across firms. DD estimators are typically used to recover the treatment effects of natural or quasi-natural experiments that trigger sharp changes in the environment of a specific group. Instead of looking at differences in just one group (e.g., the effect in the treated group), DD investigates the treatment effects by looking at the difference between differences in two groups. Such experiments are usually exploited to address endogeneity concerns (e.g., Roberts and Whited, 2013). The identifying assumption is that the outcome variable would change equally in both groups without the treatment. This assumption is also often referred to as the assumption of parallel trends. Moreover, we would ideally also want a random assignment to the treatment and control groups. Due to lobbying or other activities, this randomness is often violated in (financial) economics.

In the context of our setting, we investigate the impact of the Paris Agreement (PA), signed on December 12, 2015, on the bond yields of polluting firms. We first estimate the treatment effect of the agreement using panel regression techniques that we discuss in the Chapter 12. We then present two methods to illustrate the treatment effect over time graphically. Although we demonstrate that the treatment effect of the agreement is anticipated by bond market participants well in advance, the techniques we present below can also be applied to many other settings.

The approach we use here replicates the results of Seltzer et al. (2022) partly. Specifically, we borrow their industry definitions for grouping firms into green and brown types. Overall, the literature on ESG effects in corporate bond markets is already large but continues to grow (for recent examples, see, e.g., Halling et al. (2021), Handler et al. (2022), Huynh and Xia (2021), among many others).

The current chapter relies on this set of packages.

```
library(tidyverse)
library(lubridate)
library(RSQLite)
library(fixest)
library(broom)
```

DOI: 10.1201/b23237-13

13.1 Data Preparation

We use TRACE and Mergent FISD as data sources from our `SQLite`-database introduced in Chapters 2–4.

```
tidy_finance <- dbConnect(
  SQLite(),
  "data/tidy_finance.sqlite",
  extended_types = TRUE
)

mergent <- tbl(tidy_finance, "mergent") |>
  collect()

trace_enhanced <- tbl(tidy_finance, "trace_enhanced") |>
  collect()
```

We start our analysis by preparing the sample of bonds. We only consider bonds with a time to maturity of more than one year to the signing of the PA, so that we have sufficient data to analyze the yield behavior after the treatment date. This restriction also excludes all bonds issued after the agreement. We also consider only the first two digits of the SIC industry code to identify the polluting industries (in line with Seltzer et al., 2022).

```
treatment_date <- ymd("2015-12-12")

bonds <- mergent |>
  mutate(
    time_to_maturity = as.numeric(maturity - treatment_date),
    time_to_maturity = time_to_maturity / 365,
    sic_code = as.integer(substr(sic_code, 1, 2)),
    log_offering_amt = log(offering_amt)
  ) |>
  filter(time_to_maturity >= 1) |>
  select(
    cusip_id = complete_cusip,
    time_to_maturity, log_offering_amt, sic_code
  )

polluting_industries <- c(
  49, 13, 45, 29, 28, 33, 40, 20,
  26, 42, 10, 53, 32, 99, 37
)
```

```
bonds <- bonds |>
  mutate(polluter = sic_code %in% polluting_industries)
```

Next, we aggregate the individual transactions as reported in TRACE to a monthly panel of bond yields. We consider bond yields for a bond's last trading day in a month. Therefore, we first aggregate bond data to daily frequency and apply common restrictions from the literature (see, e.g., Bessembinder et al., 2008). We weigh each transaction by volume to reflect a trade's relative importance and avoid emphasizing small trades. Moreover, we only consider transactions with reported prices `rptd_pr` larger than 25 (to exclude bonds that are close to default) and only bond-day observations with more than five trades on a corresponding day (to exclude prices based on too few, potentially non-representative transactions).

```
trace_aggregated <- trace_enhanced |>
  filter(rptd_pr > 25) |>
  group_by(cusip_id, trd_exctn_dt) |>
  summarize(
    avg_yield = weighted.mean(yld_pt, entrd_vol_qt * rptd_pr),
    trades = n(),
    .groups = "drop"
  ) |>
  drop_na(avg_yield) |>
  filter(trades >= 5) |>
  mutate(month = floor_date(trd_exctn_dt, "months")) |>
  group_by(cusip_id, month) |>
  slice_max(trd_exctn_dt) |>
  ungroup() |>
  select(cusip_id, month, avg_yield)
```

By combining the bond-specific information from Mergent FISD for our bond sample with the aggregated TRACE data, we arrive at the main sample for our analysis.

```
bonds_panel <- bonds |>
  inner_join(trace_aggregated, by = "cusip_id") |>
  drop_na()
```

Before we can run the first regression, we need to define the `treated` indicator, which is the product of the `post_period` (i.e., all months after the signing of the PA) and the `polluter` indicator defined above.

```
bonds_panel <- bonds_panel |>
  mutate(post_period = month >= floor_date(treatment_date, "months"))
```

```
bonds_panel <- bonds_panel |>
  mutate(treated = polluter & post_period)
```

As usual, we tabulate summary statistics of the variables that enter the regression to check the validity of our variable definitions.

```
bonds_panel |>
  pivot_longer(
    cols = c(avg_yield, time_to_maturity, log_offering_amt),
    names_to = "measure"
  ) |>
  group_by(measure) |>
  summarize(
    mean = mean(value),
    sd = sd(value),
    min = min(value),
    q05 = quantile(value, 0.05),
    q50 = quantile(value, 0.50),
    q95 = quantile(value, 0.95),
    max = max(value),
    n = n(),
    .groups = "drop"
  )
```

```
# A tibble: 3 x 9
  measure           mean    sd    min   q05   q50   q95   max      n
  <chr>            <dbl> <dbl>  <dbl> <dbl> <dbl> <dbl> <dbl>  <int>
1 avg_yield         4.08 4.21  0.0595  1.27  3.37  8.07 128.  127428
2 log_offering_amt 13.3  0.823 4.64   12.2  13.2  14.5   16.5 127428
3 time_to_maturity  8.54 8.41  1.01    1.50  5.81 27.4  101.  127428
```

13.2 Panel Regressions

The PA is a legally binding international treaty on climate change. It was adopted by 196 Parties at COP 21 in Paris on 12 December 2015 and entered into force on 4 November 2016. The PA obliges developed countries to support efforts to build clean, climate-resilient futures. One may thus hypothesize that adopting climate-related policies may affect financial markets. To measure the magnitude of this effect, we first run an OLS regression without fixed effects where we include the `treated`, `post_period`, and `polluter` dummies, as well as the bond-specific characteristics `log_offering_amt` and `time_to_maturity`. This simple model assumes that there are essentially two periods (before and after the PA) and two groups (polluters and non-polluters). Nonetheless, it should indicate whether polluters have higher yields following the PA compared to non-polluters.

The second model follows the typical DD regression approach by including individual (`cusip_id`) and time (`month`) fixed effects. In this model, we do not include any other variables from the simple model because the fixed effects subsume them, and we observe the coefficient of our main variable of interest: `treated`.

```
model_without_fe <- feols(
  fml = avg_yield ~ treated + post_period + polluter +
    log_offering_amt + time_to_maturity,
  vcov = "iid",
  data = bonds_panel
)

model_with_fe <- feols(
  fml = avg_yield ~ treated | cusip_id + month,
  vcov = "iid",
  data = bonds_panel
)

etable(model_without_fe, model_with_fe, coefstat = "tstat")
```

```
                      model_without_fe     model_with_fe
Dependent Var.:              avg_yield         avg_yield

(Intercept)          10.66*** (56.60)
treatedTRUE         0.4610*** (9.284) 0.9807*** (29.43)
post_periodTRUE  -0.1747*** (-5.940)
polluterTRUE        0.4745*** (15.05)
log_offering_amt -0.5451*** (-38.55)
time_to_maturity    0.0573*** (41.29)
Fixed-Effects:   ------------------- -----------------
cusip_id                          No               Yes
month                             No               Yes
________________ ___________________ _________________
VCOV type                        IID               IID
Observations                 127,428           127,428
R2                           0.03151           0.64689
Within R2                         --           0.00713
---
Signif. codes: 0 '***' 0.001 '**' 0.01 '*' 0.05 '.' 0.1 ' ' 1
```

Both models indicate that polluters have significantly higher yields after the PA than non-polluting firms. Note that the magnitude of the `treated` coefficient varies considerably across models.

13.3 Visualizing Parallel Trends

Even though the regressions above indicate that there is an impact of the PA on bond yields of polluters, the tables do not tell us anything about the dynamics of the treatment effect. In particular, the models provide no indication about whether the crucial *parallel trends* assumption is valid. This assumption requires that in the absence of treatment, the difference between the two groups is constant over time. Although there is no well-defined statistical test for this assumption, visual inspection typically provides a good indication.

To provide such visual evidence, we revisit the simple OLS model and replace the `treated` and `post_period` indicators with month dummies for each group. This approach estimates the average yield change of both groups for each period and provides corresponding confidence intervals. Plotting the coefficient estimates for both groups around the treatment date shows us the dynamics of our panel data.

```
model_without_fe_time <- feols(
  fml = avg_yield ~ polluter + month:polluter +
    time_to_maturity + log_offering_amt,
  vcov = "iid",
  data = bonds_panel |>
    mutate(month = factor(month))
)

model_without_fe_coefs <- tidy(model_without_fe_time) |>
  filter(str_detect(term, "month")) |>
  mutate(
    month = ymd(substr(term, nchar(term) - 9, nchar(term))),
    treatment = str_detect(term, "TRUE"),
    ci_up = estimate + qnorm(0.975) * std.error,
    ci_low = estimate + qnorm(0.025) * std.error
  )

model_without_fe_coefs |>
  ggplot(aes(
    month,
    color = treatment,
    linetype = treatment,
    shape = treatment
```

```
    )) +
  geom_vline(aes(xintercept = floor_date(treatment_date, "month")),
    linetype = "dashed"
  ) +
  geom_hline(aes(yintercept = 0),
    linetype = "dashed"
  ) +
  geom_errorbar(aes(ymin = ci_low, ymax = ci_up),
    alpha = 0.5
  ) +
  guides(linetype = "none") +
  geom_point(aes(y = estimate)) +
  labs(
    x = NULL,
    y = "Yield",
    shape = "Polluter?",
    color = "Polluter?",
    title = "Polluters respond stronger to Paris Agreement than green firms"
  )
```

Figure 13.1 shows that throughout most of 2014, the yields of the two groups changed in unison. However, starting at the end of 2014, the yields start to diverge, reaching the highest difference around the signing of the PA. Afterward, the yields for both groups fall again, and the polluters arrive at the same level as at the beginning of 2014. The non-polluters, on the other hand, even experience significantly lower yields than polluters after the signing of the agreement.

Instead of plotting both groups using the simple model approach, we can also use the fixed-effects model and focus on the polluter's yield response to the signing relative to the non-polluters. To perform this estimation, we need to replace the `treated` indicator with separate time dummies for the polluters, each marking a one-month period relative to the treatment date. We then regress the monthly yields on the set of time dummies and `cusip_id` and `month` fixed effects.

```
bonds_panel_alt <- bonds_panel |>
  mutate(
    diff_to_treatment = interval(
      floor_date(treatment_date, "month"), month
    ) %/% months(1)
  )

variables <- bonds_panel_alt |>
  distinct(diff_to_treatment, month) |>
  arrange(month) |>
  mutate(variable_name = as.character(NA))
```

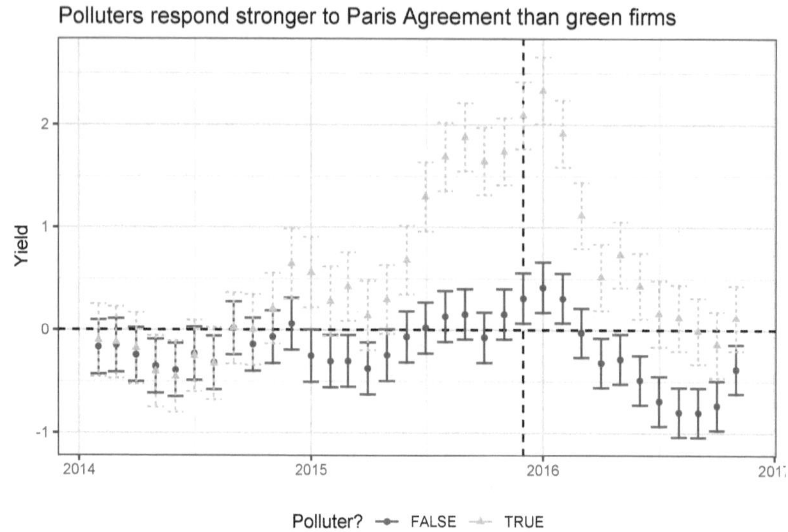

FIGURE 13.1
The figure shows the coefficient estimates and 95 percent confidence intervals for OLS regressions estimating the treatment effect of the Paris Climate Agreement on bond yields (in percent) for polluters and non-polluters. The horizontal line represents the benchmark yield of polluters before the Paris Agreement. The vertical line indicates the date of the agreement (December 12, 2015).

```
formula <- "avg_yield ~ "

for (j in 1:nrow(variables)) {
  if (variables$diff_to_treatment[j] != 0) {
    old_names <- names(bonds_panel_alt)
    bonds_panel_alt <- bonds_panel_alt |>
      mutate(new_var = diff_to_treatment == variables$diff_to_treatment[j] &
               polluter)
    new_var_name <- ifelse(variables$diff_to_treatment[j] < 0,
      str_c("lag", abs(variables$diff_to_treatment[j])),
      str_c("lead", variables$diff_to_treatment[j])
    )
    variables$variable_name[j] <- new_var_name
    names(bonds_panel_alt) <- c(old_names, new_var_name)
    formula <- str_c(
      formula,
      ifelse(j == 1,
        new_var_name,
        str_c("+", new_var_name)
```

```
      )
    )
  }
}
formula <- str_c(formula, "| cusip_id + month")

model_with_fe_time <- feols(
  fml = as.formula(formula),
  vcov = "iid",
  data = bonds_panel_alt
)

model_with_fe_time_coefs <- tidy(model_with_fe_time) |>
  mutate(
    term = str_remove(term, "TRUE"),
    ci_up = estimate + qnorm(0.975) * std.error,
    ci_low = estimate + qnorm(0.025) * std.error
  ) |>
  left_join(
    variables,
    by = c("term" = "variable_name")
  ) |>
  bind_rows(tibble(
    term = "lag0",
    estimate = 0,
    ci_up = 0,
    ci_low = 0,
    month = floor_date(treatment_date, "month")
  ))

model_with_fe_time_coefs |>
  ggplot(aes(x = month, y = estimate)) +
  geom_vline(aes(xintercept = floor_date(treatment_date, "month")),
    linetype = "dashed"
  ) +
  geom_hline(aes(yintercept = 0),
    linetype = "dashed"
  ) +
  geom_errorbar(aes(ymin = ci_low, ymax = ci_up),
    alpha = 0.5
  ) +
  geom_point(aes(y = estimate)) +
  labs(
    x = NULL,
```

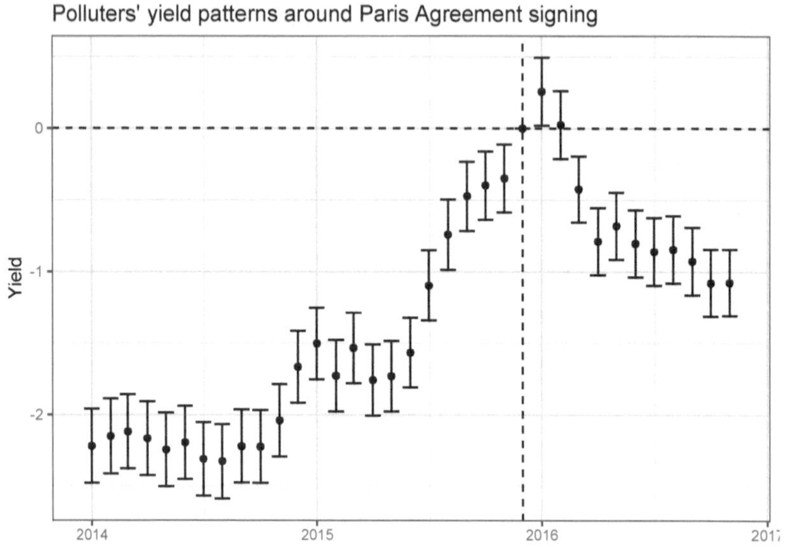

FIGURE 13.2
The figure shows the coefficient estimates and 95 percent confidence intervals for OLS regressions estimating the treatment effect of the Paris Climate Agreement on bond yields (in percent) for polluters. The horizontal line represents the benchmark yield of polluters before the Paris Agreement. The vertical line indicates the date of the agreement (December 12, 2015).

```
    y = "Yield",
    title = "Polluters' yield patterns around Paris Agreement signing"
  )
```

The resulting graph shown in Figure 13.2 confirms the main conclusion of the previous image: polluters' yield patterns show a considerable anticipation effect starting toward the end of 2014. Yields only marginally increase after the signing of the agreement. However, as opposed to the simple model, we do not see a complete reversal back to the pre-agreement level. Yields of polluters stay at a significantly higher level even one year after the signing.

Notice that during the year after the PA was signed, the 45th President of the United States was elected on November 8, 2016. During his campaign there were some indications of intentions to withdraw the US from the PA, which ultimately happened on November 4, 2020. Hence, reversal effects are potentially driven by these actions.

13.4 Exercises

1. The 46th President of the US rejoined the Paris Agreement in February 2021. Repeat the difference in differences analysis for the day of his election victory. Note that you will also have to download new TRACE data. How did polluters' yields react to this action?
2. Based on the exercise on ratings in Chapter 4, include ratings as a control variable in the analysis above. Do the results change?

14

Factor Selection via Machine Learning

The aim of this chapter is twofold. From a data science perspective, we introduce `tidymodels`, a collection of packages for modeling and machine learning (ML) using `tidyverse` principles. `tidymodels` comes with a handy workflow for all sorts of typical prediction tasks. From a finance perspective, we address the notion of *factor zoo* (Cochrane, 2011) using ML methods. We introduce Lasso and Ridge regression as a special case of penalized regression models. Then, we explain the concept of cross-validation for model *tuning* with Elastic Net regularization as a popular example. We implement and showcase the entire cycle from model specification, training, and forecast evaluation within the `tidymodels` universe. While the tools can generally be applied to an abundance of interesting asset pricing problems, we apply penalized regressions for identifying macroeconomic variables and asset pricing factors that help explain a cross-section of industry portfolios.

In previous chapters, we illustrate that stock characteristics such as size provide valuable pricing information in addition to the market beta. Such findings question the usefulness of the Capital Asset Pricing Model. In fact, during the last decades, financial economists discovered a plethora of additional factors which may be correlated with the marginal utility of consumption (and would thus deserve a prominent role in pricing applications). The search for factors that explain the cross-section of expected stock returns has produced hundreds of potential candidates, as noted more recently by Harvey et al. (2016), Mclean and Pontiff (2016), and Hou et al. (2020). Therefore, given the multitude of proposed risk factors, the challenge these days rather is: *do we believe in the relevance of 300+ risk factors?* During recent years, promising methods from the field of ML got applied to common finance applications. We refer to Mullainathan and Spiess (2017) for a treatment of ML from the perspective of an econometrician, Nagel (2021) for an excellent review of ML practices in asset pricing, Easley et al. (2020) for ML applications in (high-frequency) market microstructure, and Dixon et al. (2020) for a detailed treatment of all methodological aspects.

DOI: 10.1201/b23237-14

14.1 Brief Theoretical Background

This is a book about *doing* empirical work in a tidy manner, and we refer to any of the many excellent textbook treatments of ML methods and especially penalized regressions for some deeper discussion. Excellent material is provided, for instance, by Hastie et al. (2009), Gareth et al. (2013), and De Prado (2018). Instead, we briefly summarize the idea of Lasso and Ridge regressions as well as the more general Elastic Net. Then, we turn to the fascinating question on *how* to implement, tune, and use such models with the `tidymodels` workflow.

To set the stage, we start with the definition of a linear model: suppose we have data $(y_t, x_t), t = 1, \ldots, T$, where x_t is a $(K \times 1)$ vector of regressors and y_t is the response for observation t. The linear model takes the form $y_t = \beta' x_t + \varepsilon_t$ with some error term ε_t and has been studied in abundance. The well-known ordinary-least square (OLS) estimator for the $(K \times 1)$ vector β minimizes the sum of squared residuals and is then

$$\hat{\beta}^{\text{ols}} = \left(\sum_{t=1}^{T} x_t' x_t\right)^{-1} \sum_{t=1}^{T} x_t' y_t.$$

While we are often interested in the estimated coefficient vector $\hat{\beta}^{\text{ols}}$, ML is about the predictive performance most of the time. For a new observation $\tilde{x}_t$, the linear model generates predictions such that

$$\hat{y}_t = E\left(y | x_t = \tilde{x}_t\right) = \hat{\beta}^{\text{ols}\prime} \tilde{x}_t.$$

Is this the best we can do? Not really: instead of minimizing the sum of squared residuals, penalized linear models can improve predictive performance by choosing other estimators $\hat{\beta}$ with lower variance than the estimator $\hat{\beta}^{\text{ols}}$. At the same time, it seems appealing to restrict the set of regressors to a few meaningful ones if possible. In other words, if K is large (such as for the number of proposed factors in the asset pricing literature), it may be a desirable feature to *select* reasonable factors and set $\hat{\beta}^{\text{ols}}_k = 0$ for some redundant factors.

It should be clear that the promised benefits of penalized regressions, i.e., reducing the mean squared error (MSE), come at a cost. In most cases, reducing the variance of the estimator introduces a bias such that $E\left(\hat{\beta}\right) \neq \beta$. What is the effect of such a bias-variance trade-off? To understand the implications, assume the following data-generating process for y:

$$y = f(x) + \varepsilon, \quad \varepsilon \sim (0, \sigma_\varepsilon^2)$$

We want to recover $f(x)$, which denotes some unknown functional which maps the relationship between x and y. While the properties of $\hat{\beta}^{\text{ols}}$ as an unbiased estimator may be desirable under some circumstances, they are certainly not if we consider predictive accuracy. Alternative predictors $\hat{f}(x)$ could be more desirable: For instance, the MSE depends on our model choice as follows:

$$\begin{aligned}
MSE &= E((y - \hat{f}(x))^2) = E((f(x) + \epsilon - \hat{f}(x))^2) \\
&= \underbrace{E((f(x) - \hat{f}(x))^2)}_{\text{total quadratic error}} + \underbrace{E(\epsilon^2)}_{\text{irreducible error}} \\
&= E\left(\hat{f}(x)^2\right) + E\left(f(x)^2\right) - 2E\left(f(x)\hat{f}(x)\right) + \sigma_\varepsilon^2 \\
&= E\left(\hat{f}(x)^2\right) + f(x)^2 - 2f(x)E\left(\hat{f}(x)\right) + \sigma_\varepsilon^2 \\
&= \underbrace{\text{Var}\left(\hat{f}(x)\right)}_{\text{variance of model}} + \underbrace{E\left((f(x) - \hat{f}(x))\right)^2}_{\text{squared bias}} + \sigma_\varepsilon^2.
\end{aligned}$$

While no model can reduce σ_ε^2, a biased estimator with small variance may have a lower MSE than an unbiased estimator.

14.1.1 Ridge regression

One biased estimator is known as Ridge regression. Hoerl and Kennard (1970) propose to minimize the sum of squared errors *while simultaneously imposing a penalty on the L_2 norm of the parameters* $\hat{\beta}$. Formally, this means that for a penalty factor $\lambda \geq 0$ the minimization problem takes the form $\min_\beta (y - X\beta)'(y - X\beta)$ s.t. $\beta'\beta \leq c$. Here $c \geq 0$ is a constant that depends on the choice of λ. The larger λ, the smaller c (technically speaking, there is a one-to-one relationship between λ, which corresponds to the Lagrangian of the minimization problem above and c). Here, $X = (x_1 \ldots x_T)'$ and $y = (y_1, \ldots, y_T)'$. A closed-form solution for the resulting regression coefficient vector β^{ridge} exists:

$$\hat{\beta}^{\text{ridge}} = (X'X + \lambda I)^{-1} X'y.$$

A couple of observations are worth noting: $\hat{\beta}^{\text{ridge}} = \hat{\beta}^{\text{ols}}$ for $\lambda = 0$ and $\hat{\beta}^{\text{ridge}} \to 0$ for $\lambda \to \infty$. Also for $\lambda > 0$, $(X'X + \lambda I)$ is non-singular even if $X'X$ is which means that $\hat{\beta}^{\text{ridge}}$ exists even if $\hat{\beta}$ is not defined. However, note also that the Ridge estimator requires careful choice of the hyperparameter λ which controls the *amount of regularization*: a larger value of λ implies *shrinkage* of the regression coefficient toward 0, a smaller value of λ reduces the bias of the resulting estimator.

Note, that X usually contains an intercept column with ones. As a general rule, the associated intercept coefficient is not penalized. In practice, this often implies that y is simply demeaned before computing $\hat{\beta}^{\text{ridge}}$.

What about the statistical properties of the Ridge estimator? First, the bad news is that $\hat{\beta}^{\text{ridge}}$ is a biased estimator of β. However, the good news is that (under homoscedastic error terms) the variance of the Ridge estimator is guaranteed to be *smaller* than the variance of the ordinary least square estimator. We encourage you to verify these two statements in the exercises. As a result, we face a trade-off: The Ridge regression sacrifices some unbiasedness to achieve a smaller variance than the OLS estimator.

14.1.2 Lasso

An alternative to Ridge regression is the Lasso (*l*east *a*bsolute *s*hrinkage and *s*election *o*perator). Similar to Ridge regression, the Lasso (Tibshirani, 1996) is a penalized and biased estimator. The main difference to Ridge regression is that Lasso does not only *shrink* coefficients but effectively selects variables by setting coefficients for *irrelevant* variables to zero. Lasso implements a L_1 penalization on the parameters such that:

$$\hat{\beta}^{\text{Lasso}} = \arg\min_{\beta} \left(Y - X\beta\right)'\left(Y - X\beta\right) \text{ s.t. } \sum_{k=1}^{K} |\beta_k| < c(\lambda).$$

There is no closed form solution for $\hat{\beta}^{\text{Lasso}}$ in the above maximization problem but efficient algorithms exist (e.g., the R package `glmnet`). Like for Ridge regression, the hyperparameter λ has to be specified beforehand.

14.1.3 Elastic Net

The Elastic Net (Zou and Hastie, 2005) combines L_1 with L_2 penalization and encourages a grouping effect, where strongly correlated predictors tend to be in or out of the model together. This more general framework considers the following optimization problem:

$$\hat{\beta}^{\text{EN}} = \arg\min_{\beta} \left(Y - X\beta\right)'\left(Y - X\beta\right) + \lambda(1-\rho)\sum_{k=1}^{K}|\beta_k| + \frac{1}{2}\lambda\rho\sum_{k=1}^{K}\beta_k^2$$

Now, we have to chose two hyperparameters: the *shrinkage* factor λ and the *weighting parameter* ρ. The Elastic Net resembles Lasso for $\rho = 1$ and Ridge regression for $\rho = 0$. While the R package `glmnet` provides efficient algorithms to compute the coefficients of penalized regressions, it is a good exercise to implement Ridge and Lasso estimation on your own before you use the `glmnet` package or the `tidymodels` back-end.

14.2 Data Preparation

To get started, we load the required packages and data. The main focus is on the workflow behind the `tidymodels` package collection (Kuhn and Wickham, 2020). Kuhn and Silge (2018) provide a thorough introduction into all `tidymodels` components. `glmnet` (Simon et al., 2011) was developed and released in sync with Tibshirani (1996) and provides an R implementation of Elastic Net estimation. The package `timetk` (Dancho and Vaughan, 2022b) provides useful tools for time series data wrangling.

```
library(RSQLite)
library(tidyverse)
library(scales)
library(furrr)
library(broom)
library(tidymodels)
library(glmnet)
library(timetk)
```

In this analysis, we use four different data sources that we load from our `SQLite`-database introduced in Chapters 2–4. We start with two different sets of factor portfolio returns which have been suggested as representing practical risk factor exposure and thus should be relevant when it comes to asset pricing applications.

- The standard workhorse: monthly Fama-French 3 factor returns (market, small-minus-big, and high-minus-low book-to-market valuation sorts) defined in Fama and French (1992) and Fama and French (1993)
- Monthly q-factor returns from Hou et al. (2014). The factors contain the size factor, the investment factor, the return-on-equity factor, and the expected growth factor

Next, we include macroeconomic predictors which may predict the general stock market economy. Macroeconomic variables effectively serve as conditioning information such that their inclusion hints at the relevance of conditional models instead of unconditional asset pricing. We refer the interested reader to Cochrane (2009) on the role of conditioning information.

- Our set of macroeconomic predictors comes from Welch and Goyal (2008). The data has been updated by the authors until 2021 and contains monthly variables that have been suggested as good predictors for the equity premium. Some of the variables are the dividend price ratio, earnings price ratio, stock variance, net equity expansion, treasury bill rate, and inflation

Finally, we need a set of *test assets*. The aim is to understand which of the plenty factors and macroeconomic variable combinations prove helpful in explaining our test assets' cross-section of returns. In line with many existing papers, we use monthly portfolio returns from 10 different industries according to the definition from Kenneth French's homepage[1] as test assets.

```
tidy_finance <- dbConnect(
  SQLite(),
  "data/tidy_finance.sqlite",
  extended_types = TRUE
)

factors_ff_monthly <- tbl(tidy_finance, "factors_ff_monthly") |>
  collect() |>
  rename_with(~ str_c("factor_ff_", .), -month)

factors_q_monthly <- tbl(tidy_finance, "factors_q_monthly") |>
  collect() |>
  rename_with(~ str_c("factor_q_", .), -month)

macro_predictors <- tbl(tidy_finance, "macro_predictors") |>
  collect() |>
  rename_with(~ str_c("macro_", .), -month) |>
  select(-macro_rp_div)

industries_ff_monthly <- tbl(tidy_finance, "industries_ff_monthly") |>
  collect() |>
  pivot_longer(-month,
    names_to = "industry", values_to = "ret"
  ) |>
  mutate(industry = as_factor(industry))
```

We combine all the monthly observations into one data frame.

```
data <- industries_ff_monthly |>
  left_join(factors_ff_monthly, by = "month") |>
  left_join(factors_q_monthly, by = "month") |>
  left_join(macro_predictors, by = "month") |>
  mutate(
    ret = ret - factor_ff_rf
  ) |>
  select(month, industry, ret, everything()) |>
  drop_na()
```

[1] https://mba.tuck.dartmouth.edu/pages/faculty/ken.french/Data_Library/det_10_ind_port.html

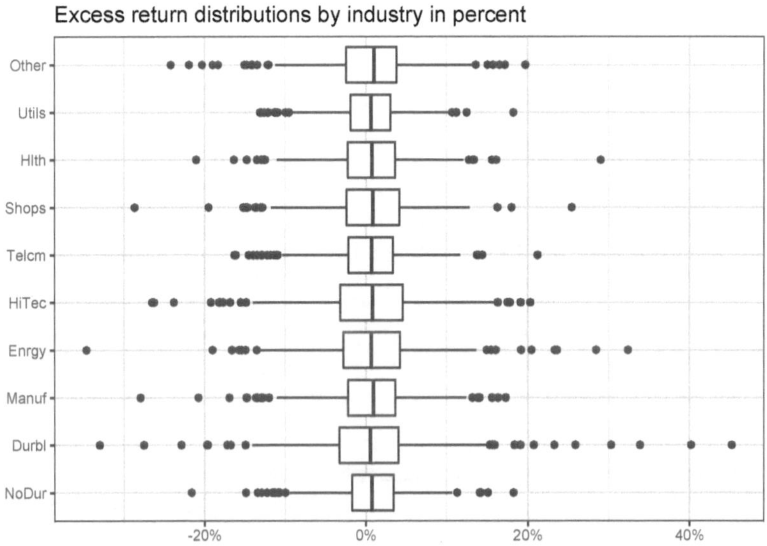

FIGURE 14.1
The box plots show the monthly dispersion of returns for 10 different industries.

Our data contains 22 columns of regressors with the 13 macro variables and 8 factor returns for each month. Figure 14.1 provides summary statistics for the 10 monthly industry excess returns in percent.

```
data |>
  group_by(industry) |>
  mutate(ret = ret) |>
  ggplot(aes(x = industry, y = ret)) +
  geom_boxplot() +
  coord_flip() +
  labs(
    x = NULL, y = NULL,
    title = "Excess return distributions by industry in percent"
  ) +
  scale_y_continuous(
    labels = percent
  )
```

14.3 The Tidymodels Workflow

To illustrate penalized linear regressions, we employ the `tidymodels` collection of packages for modeling and ML using `tidyverse` principles. You can simply use `install.packages("tidymodels")` to get access to all the related packages. We recommend checking out the work of Kuhn and Silge (2018): They continuously write on their great book 'Tidy Modeling with R'[2] using tidy principles.

The `tidymodels` workflow encompasses the main stages of the modeling process: pre-processing of data, model fitting, and post-processing of results. As we demonstrate below, `tidymodels` provides efficient workflows that you can update with low effort.

Using the ideas of Ridge and Lasso regressions, the following example guides you through (i) pre-processing the data (data split and variable mutation), (ii) building models, (iii) fitting models, and (iv) tuning models to create the "best" possible predictions.

To start, we restrict our analysis to just one industry: Manufacturing. We first split the sample into a *training* and a *test* set. For that purpose, `tidymodels` provides the function `initial_time_split()` from the `rsample` package (Silge et al., 2022). The split takes the last 20% of the data as a test set, which is not used for any model tuning. We use this test set to evaluate the predictive accuracy in an out-of-sample scenario.

```
split <- initial_time_split(
  data |>
    filter(industry == "Manuf") |>
    select(-industry),
  prop = 4 / 5
)
split
```

```
<Training/Testing/Total>
<527/132/659>
```

The object `split` simply keeps track of the observations of the training and the test set. We can call the training set with `training(split)`, while we can extract the test set with `testing(split)`.

14.3.1 Pre-process data

Recipes help you pre-process your data before training your model. Recipes are a series of pre-processing steps such as variable selection, transformation, or conversion

[2] https://www.tmwr.org/

of qualitative predictors to indicator variables. Each recipe starts with a `formula` that defines the general structure of the dataset and the role of each variable (regressor or dependent variable). For our dataset, our recipe contains the following steps before we fit any model:

- Our formula defines that we want to explain excess returns with all available predictors. The regression equation thus takes the form

$$r_t = \alpha_0 + \left(\tilde{f}_t \otimes \tilde{z}_t\right) B + \varepsilon_t$$

  where r_t is the vector of industry excess returns at time t and $\tilde{f}_t$ and $\tilde{z}_t$ are the (standardized) vectors of factor portfolio returns and macroeconomic variables
- We exclude the column *month* from the analysis
- We include all interaction terms between factors and macroeconomic predictors
- We demean and scale each regressor such that the standard deviation is one

```
rec <- recipe(ret ~ ., data = training(split)) |>
  step_rm(month) |>
  step_interact(terms = ~ contains("factor"):contains("macro")) |>
  step_normalize(all_predictors()) |>
  step_center(ret, skip = TRUE)
```

A table of all available recipe steps can be found in the `tidymodels` documentation.[3] As of 2022, more than 150 different processing steps are available! One important point: The definition of a recipe does not trigger any calculations yet but rather provides a *description* of the tasks to be applied. As a result, it is very easy to *reuse* recipes for different models and thus make sure that the outcomes are comparable as they are based on the same input. In the example above, it does not make a difference whether you use the input `data = training(split)` or `data = testing(split)`. All that matters at this early stage are the column names and types.

We can apply the recipe to any data with a suitable structure. The code below combines two different functions: `prep()` estimates the required parameters from a training set that can be applied to other data sets later. `bake()` applies the processed computations to new data.

```
data_prep <- prep(rec, training(split))
```

The object `data_prep` contains information related to the different preprocessing steps applied to the training data: E.g., it is necessary to compute sample means and standard deviations to center and scale the variables.

[3] https://www.tidymodels.org/find/recipes/

```
data_bake <- bake(data_prep,
  new_data = testing(split)
)
data_bake
```

```
# A tibble: 132 x 126
  factor_ff~1 facto~2 facto~3 facto~4 facto~5 facto~6 facto~7 facto~8
        <dbl>   <dbl>   <dbl>   <dbl>   <dbl>   <dbl>   <dbl>   <dbl>
1       -1.80 1.37      0.139   1.14   0.0289   1.33  -1.63    -1.66
2       -1.80 0.333    -0.833   0.121 -0.745    0.450 -0.0117  -1.19
3       -1.80 0.656     0.399   0.297  0.327    0.510 -0.860   -0.905
4       -1.80 0.00426   0.717  -0.756  0.470   -0.296 -0.590   -0.324
5       -1.84 0.529    -0.177  -0.972 -0.390   -0.733  0.383    0.102
# ... with 127 more rows, 118 more variables: macro_dp <dbl>,
#   macro_dy <dbl>, macro_ep <dbl>, macro_de <dbl>,
#   macro_svar <dbl>, macro_bm <dbl>, macro_ntis <dbl>,
#   macro_tbl <dbl>, macro_lty <dbl>, macro_ltr <dbl>,
#   macro_tms <dbl>, macro_dfy <dbl>, macro_infl <dbl>, ret <dbl>,
#   factor_ff_rf_x_macro_dp <dbl>, factor_ff_rf_x_macro_dy <dbl>,
#   factor_ff_rf_x_macro_ep <dbl>, ...
```

Note that the resulting data contains the 132 observations from the test set and 126 columns. Why so many? Recall that the recipe states to compute every possible interaction term between the factors and predictors, which increases the dimension of the data matrix substantially.

You may ask at this stage: why should I use a recipe instead of simply using the data wrangling commands such as `mutate()` or `select()`? `tidymodels` beauty is that a lot is happening under the hood. Recall, that for the simple scaling step, you actually have to compute the standard deviation of each column, then *store* this value, and apply the identical transformation to a different dataset, e.g., `testing(split)`. A prepped `recipe` stores these values and hands them on once you `bake()` a novel dataset. Easy as pie with `tidymodels`, isn't it?

14.3.2 Build a model

Next, we can build an actual model based on our pre-processed data. In line with the definition above, we estimate regression coefficients of a Lasso regression such that we get

$$\hat{\beta}_\lambda^{\text{Lasso}} = \arg\min_\beta \left(Y - X\beta\right)'\left(Y - X\beta\right) + \lambda\sum_{k=1}^{K}|\beta_k|.$$

We want to emphasize that the `tidymodels` workflow for *any* model is very similar, irrespective of the specific model. As you will see further below, it is straightforward to fit Ridge regression coefficients and – later – Neural networks or Random forests with basically the same code. The structure is always as follows: create a so-called `workflow()` and use the `fit()` function. A table with all available model APIs is available here.[4] For now, we start with the linear regression model with a given value for the penalty factor λ. In the setup below, `mixture` denotes the value of ρ, hence setting `mixture = 1` implies the Lasso.

```
lm_model <- linear_reg(
  penalty = 0.0001,
  mixture = 1
) |>
  set_engine("glmnet", intercept = FALSE)
```

That's it – we are done! The object `lm_model` contains the definition of our model with all required information. Note that `set_engine("glmnet")` indicates the API character of the `tidymodels` workflow: Under the hood, the package `glmnet` is doing the heavy lifting, while `linear_reg()` provides a unified framework to collect the inputs. The `workflow` ends with combining everything necessary for the serious data science workflow, namely, a recipe and a model.

```
lm_fit <- workflow() |>
  add_recipe(rec) |>
  add_model(lm_model)
lm_fit
```

```
== Workflow =========================================================
Preprocessor: Recipe
Model: linear_reg()

-- Preprocessor -----------------------------------------------------
4 Recipe Steps

* step_rm()
* step_interact()
* step_normalize()
* step_center()

-- Model ------------------------------------------------------------
Linear Regression Model Specification (regression)

Main Arguments:
  penalty = 1e-04
  mixture = 1
```

[4] https://www.tidymodels.org/find/parsnip/

```
Engine-Specific Arguments:
  intercept = FALSE

Computational engine: glmnet
```

14.3.3 Fit a model

With the `workflow` from above, we are ready to use `fit()`. Typically, we use training data to fit the model. The training data is pre-processed according to our recipe steps, and the Lasso regression coefficients are computed. First, we focus on the predicted values $\hat{y}_t = x_t\hat{\beta}^{\text{Lasso}}$. Figure 14.2 illustrates the projections for the *entire* time series of the manufacturing industry portfolio returns. The grey area indicates the out-of-sample period, which we did not use to fit the model.

```
predicted_values <- lm_fit |>
  fit(data = training(split)) |>
  predict(data |> filter(industry == "Manuf")) |>
  bind_cols(data |> filter(industry == "Manuf")) |>
  select(month,
    "Fitted value" = .pred,
    "Realization" = ret
  ) |>
  pivot_longer(-month, names_to = "Variable")
```

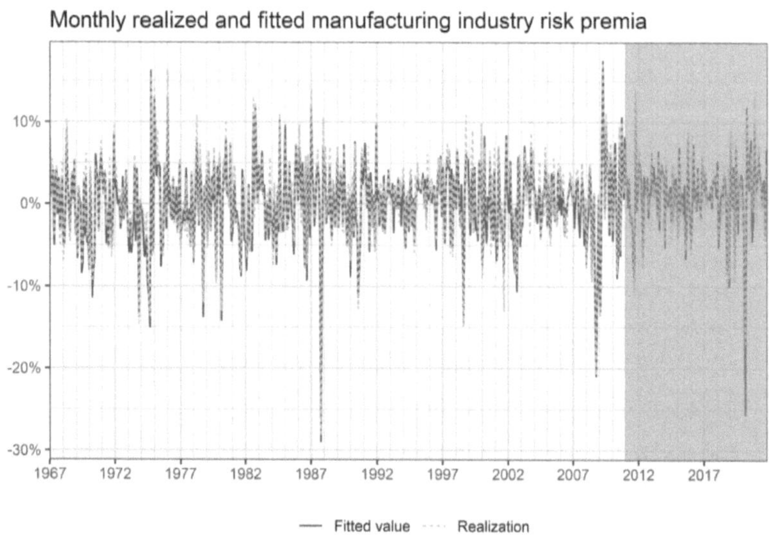

FIGURE 14.2
The grey area corresponds to the out of sample period.

```
predicted_values |>
  ggplot(aes(
    x = month,
    y = value,
    color = Variable,
    linetype = Variable
    )) +
  geom_line() +
  labs(
    x = NULL,
    y = NULL,
    color = NULL,
    linetype = NULL,
    title = "Monthly realized and fitted manufacturing industry risk premia"
  ) +
  scale_x_date(
    breaks = function(x) {
      seq.Date(
        from = min(x),
        to = max(x),
        by = "5 years"
      )
    },
    minor_breaks = function(x) {
      seq.Date(
        from = min(x),
        to = max(x),
        by = "1 years"
      )
    },
    expand = c(0, 0),
    labels = date_format("%Y")
  ) +
  scale_y_continuous(
    labels = percent
  ) +
  annotate("rect",
    xmin = testing(split) |> pull(month) |> min(),
    xmax = testing(split) |> pull(month) |> max(),
    ymin = -Inf, ymax = Inf,
    alpha = 0.5, fill = "grey70"
  )
```

What do the estimated coefficients look like? To analyze these values and to illustrate the difference between the `tidymodels` workflow and the underlying `glmnet` package, it is worth computing the coefficients $\hat{\beta}^{\text{Lasso}}$ directly. The code below estimates the coefficients for the Lasso and Ridge regression for the processed training data sample. Note that `glmnet` actually takes a vector `y` and the matrix of regressors X as input. Moreover, `glmnet` requires choosing the penalty parameter α, which corresponds to ρ in the notation above. When using the `tidymodels` model API, such details do not need consideration.

```
x <- data_bake |>
  select(-ret) |>
  as.matrix()
y <- data_bake |> pull(ret)

fit_lasso <- glmnet(
  x = x,
  y = y,
  alpha = 1,
  intercept = FALSE,
  standardize = FALSE,
  lambda.min.ratio = 0
)

fit_ridge <- glmnet(
  x = x,
  y = y,
  alpha = 0,
  intercept = FALSE,
  standardize = FALSE,
  lambda.min.ratio = 0
)
```

The objects `fit_lasso` and `fit_ridge` contain an entire sequence of estimated coefficients for multiple values of the penalty factor λ. Figure 14.3 illustrates the trajectories of the regression coefficients as a function of the penalty factor. Both Lasso and Ridge coefficients converge to zero as the penalty factor increases.

```
bind_rows(
  tidy(fit_lasso) |> mutate(Model = "Lasso"),
  tidy(fit_ridge) |> mutate(Model = "Ridge")
) |>
  rename("Variable" = term) |>
  ggplot(aes(x = lambda, y = estimate, color = Variable)) +
  geom_line() +
  scale_x_log10() +
```

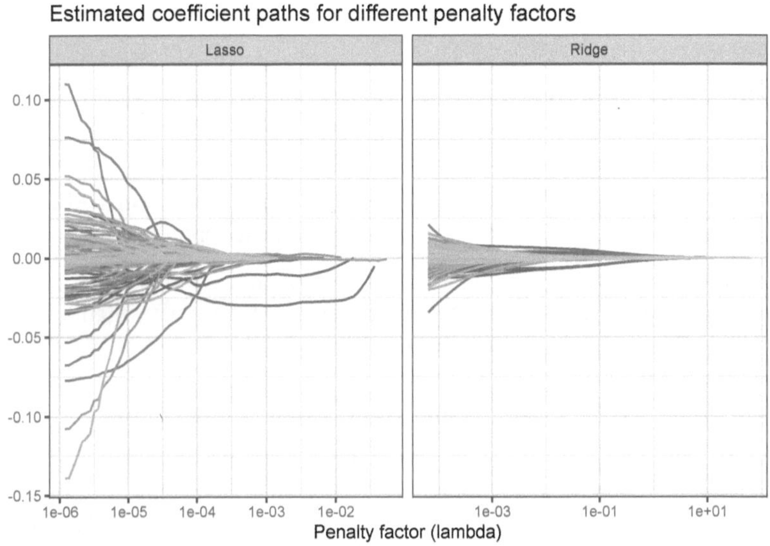

FIGURE 14.3
The penalty parameters are chosen iteratively to resemble the path from no penalization to a model that excludes all variables.

```
facet_wrap(~Model, scales = "free_x") +
labs(
  x = "Penalty factor (lambda)", y = NULL,
  title = "Estimated coefficient paths for different penalty factors"
) +
theme(legend.position = "none")
```

One word of caution: The package `glmnet` computes estimates of the coefficients $\hat{\beta}$ based on numerical optimization procedures. As a result, the estimated coefficients for the special case[5] with no regularization ($\lambda = 0$) can deviate from the standard OLS estimates.

14.3.4 Tune a model

To compute $\hat{\beta}_{\lambda}^{\text{Lasso}}$, we simply imposed a value for the penalty hyperparameter λ. Model tuning is the process of optimally selecting such hyperparameters. `tidymodels` provides extensive tuning options based on so-called *cross-validation*. Again, we refer to any treatment of cross-validation to get a more detailed discussion of the statistical

[5] https://parsnip.tidymodels.org/reference/glmnet-details.html

underpinnings. Here we focus on the general idea and the implementation with `tidymodels`.

The goal for choosing λ (or any other hyperparameter, e.g., ρ for the Elastic Net) is to find a way to produce predictors $\hat{Y}$ for an outcome Y that minimizes the mean squared prediction error $\text{MSPE} = E\left(\frac{1}{T}\sum_{t=1}^{T}(\hat{y}_t - y_t)^2\right)$. Unfortunately, the MSPE is not directly observable. We can only compute an estimate because our data is random and because we do not observe the entire population.

Obviously, if we train an algorithm on the same data that we use to compute the error, our estimate $\hat{\text{MSPE}}$ would indicate way better predictive accuracy than what we can expect in real out-of-sample data. The result is called overfitting.

Cross-validation is a technique that allows us to alleviate this problem. We approximate the true MSPE as the average of many MSPE obtained by creating predictions for K new random samples of the data, none of them used to train the algorithm $\frac{1}{K}\sum_{k=1}^{K}\frac{1}{T}\sum_{t=1}^{T}\left(\hat{y}_t^k - y_t^k\right)^2$. In practice, this is done by carving out a piece of our data and pretending it is an independent sample. We again divide the data into a training set and a test set. The MSPE on the test set is our measure for actual predictive ability, while we use the training set to fit models with the aim to find the *optimal* hyperparameter values. To do so, we further divide our training sample into (several) subsets, fit our model for a grid of potential hyperparameter values (e.g., λ), and evaluate the predictive accuracy on an *independent* sample. This works as follows:

1. Specify a grid of hyperparameters
2. Obtain predictors $\hat{y}_i(\lambda)$ to denote the predictors for the used parameters λ
3. Compute
$$\text{MSPE}(\lambda) = \frac{1}{K}\sum_{k=1}^{K}\frac{1}{T}\sum_{t=1}^{T}\left(\hat{y}_t^k(\lambda) - y_t^k\right)^2$$
With K-fold cross-validation, we do this computation K times. Simply pick a validation set with $M = T/K$ observations at random and think of these as random samples $y_1^k, \ldots, y_{\tilde{T}}^k$, with $k = 1$

How should you pick K? Large values of K are preferable because the training data better imitates the original data. However, larger values of K will have much higher computation time. `tidymodels` provides all required tools to conduct K-fold cross-validation. We just have to update our model specification and let `tidymodels` know which parameters to tune. In our case, we specify the penalty factor λ as well as the mixing factor ρ as *free* parameters. Note that it is simple to change an existing `workflow` with `update_model()`.

```
lm_model <- linear_reg(
  penalty = tune(),
  mixture = tune()
) |>
```

```
  set_engine("glmnet")

lm_fit <- lm_fit |>
  update_model(lm_model)
```

For our sample, we consider a time-series cross-validation sample. This means that we tune our models with 20 random samples of length five years with a validation period of four years. For a grid of possible hyperparameters, we then fit the model for each fold and evaluate $\widehat{\text{MSPE}}$ in the corresponding validation set. Finally, we select the model specification with the lowest MSPE in the validation set. First, we define the cross-validation folds based on our training data only.

```
data_folds <- time_series_cv(
  data        = training(split),
  date_var    = month,
  initial     = "5 years",
  assess      = "48 months",
  cumulative  = FALSE,
  slice_limit = 20
)

data_folds
```

```
# Time Series Cross Validation Plan
# A tibble: 20 x 2
  splits          id
  <list>          <chr>
1 <split [60/48]> Slice01
2 <split [60/48]> Slice02
3 <split [60/48]> Slice03
4 <split [60/48]> Slice04
5 <split [60/48]> Slice05
# ... with 15 more rows
```

Then, we evaluate the performance for a grid of different penalty values. `tidymodels` provides functionalities to construct a suitable grid of hyperparameters with `grid_regular`. The code chunk below creates a 10×3 hyperparameters grid. Then, the function `tune_grid()` evaluates all the models for each fold.

```
lm_tune <- lm_fit |>
  tune_grid(
    resample = data_folds,
    grid = grid_regular(penalty(), mixture(), levels = c(10, 3)),
    metrics = metric_set(rmse)
  )
```

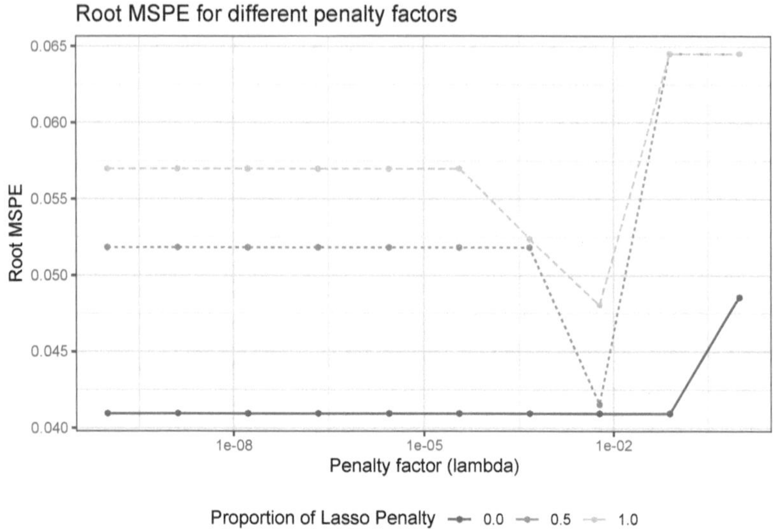

FIGURE 14.4
Evaluation of manufacturing excess returns for different penalty factors (lambda) and proportions of Lasso penalty (rho). 1.0 indicates Lasso, 0.5 indicates Elastic Net, and 0.0 indicates Ridge.

After the tuning process, we collect the evaluation metrics (the root mean-squared error in our example) to identify the *optimal* model. Figure 14.4 illustrates the average validation set's root mean-squared error for each value of λ and ρ.

```
autoplot(lm_tune) +
  aes(linetype = `Proportion of Lasso Penalty`) +
  guides(linetype = "none") +
  labs(
    x = "Penalty factor (lambda)",
    y = "Root MSPE",
    title = "Root MSPE for different penalty factors"
  )
```

Figure 14.4 shows that the cross-validated MSPE drops for Lasso and Elastic Net and spikes afterward. For Ridge regression, the MSPE increases above a certain threshold. Recall that the larger the regularization, the more restricted the model becomes. Thus, we would choose the model with the lowest MSPE.

14.3.5 Parallelized workflow

Our starting point was the question: Which factors determine industry returns? While Avramov et al. (2022b) provide a Bayesian analysis related to the research question above, we choose a simplified approach: To illustrate the entire workflow, we now run the penalized regressions for all ten industries. We want to identify relevant variables by fitting Lasso models for each industry returns time series. More specifically, we perform cross-validation for each industry to identify the optimal penalty factor λ. Then, we use the set of `finalize_*()`-functions that take a list or tibble of tuning parameter values and update objects with those values. After determining the best model, we compute the final fit on the entire training set and analyze the estimated coefficients.

First, we define the Lasso model with one tuning parameter.

```
lasso_model <- linear_reg(
  penalty = tune(),
  mixture = 1
) |>
  set_engine("glmnet")

lm_fit <- lm_fit |>
  update_model(lasso_model)
```

Figure 14.4 task can be easily parallelized to reduce computing time substantially. We use the parallelization capabilities of `furrr`. Note that we can also just recycle all the steps from above and collect them in a function.

```
select_variables <- function(input) {
  # Split into training and testing data
  split <- initial_time_split(input, prop = 4 / 5)

  # Data folds for cross-validation
  data_folds <- time_series_cv(
    data = training(split),
    date_var = month,
    initial = "5 years",
    assess = "48 months",
    cumulative = FALSE,
    slice_limit = 20
  )

  # Model tuning with the Lasso model
  lm_tune <- lm_fit |>
    tune_grid(
```

```
    resample = data_folds,
    grid = grid_regular(penalty(), levels = c(10)),
    metrics = metric_set(rmse)
  )

  # Identify the best model and fit with the training data
  lasso_lowest_rmse <- lm_tune |> select_by_one_std_err("rmse")
  lasso_final <- finalize_workflow(lm_fit, lasso_lowest_rmse)
  lasso_final_fit <- last_fit(lasso_final, split, metrics = metric_set(rmse))

  # Extract the estimated coefficients
  estimated_coefficients <- lasso_final_fit |>
    extract_fit_parsnip() |>
    tidy() |>
    mutate(
      term = str_remove_all(term, "factor_|macro_|industry_")
    )

  return(estimated_coefficients)
}

# Parallelization
plan(multisession, workers = availableCores())

# Computation by industry
selected_factors <- data |>
  nest(data = -industry) |>
  mutate(selected_variables = future_map(
    data, select_variables,
    .options = furrr_options(seed = TRUE)
  ))
```

What has just happened? In principle, exactly the same as before but instead of computing the Lasso coefficients for one industry, we did it for ten in parallel. The final option `seed = TRUE` is required to make the cross-validation process reproducible. Now, we just have to do some housekeeping and keep only variables that Lasso does *not* set to zero. We illustrate the results in a heat map in Figure 14.5

```
selected_factors |>
  unnest(selected_variables) |>
  filter(
    term != "(Intercept)",
    estimate != 0
  ) |>
  add_count(term) |>
```

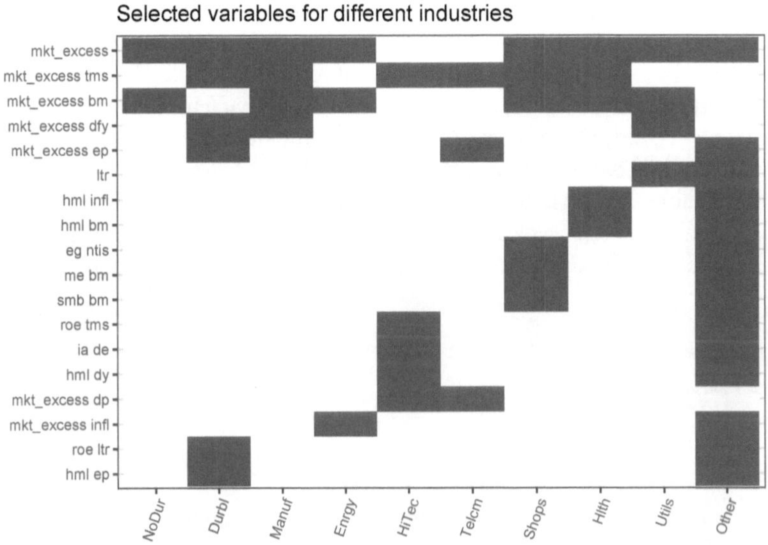

FIGURE 14.5
Grey areas indicate that the estimated Lasso regression coefficient is not set to zero. White fields show which variables get assigned a value of exactly zero.

```
mutate(
  term = str_remove_all(term, "NA|ff_|q_"),
  term = str_replace_all(term, "_x_", " "),
  term = fct_reorder(as_factor(term), n),
  term = fct_lump_min(term, min = 2),
  selected = 1
) |>
filter(term != "Other") |>
mutate(term = fct_drop(term)) |>
complete(industry, term, fill = list(selected = 0)) |>
ggplot(aes(industry,
  term,
  fill = as_factor(selected)
)) +
geom_tile() +
scale_x_discrete(guide = guide_axis(angle = 70)) +
scale_fill_manual(values = c("white", "grey30")) +
theme(legend.position = "None") +
labs(
  x = NULL, y = NULL,
  title = "Selected variables for different industries"
)
```

The heat map in Figure 14.5 conveys two main insights. First, we see a lot of white, which means that many factors, macroeconomic variables, and interaction terms are not relevant for explaining the cross-section of returns across the industry portfolios. In fact, only the market factor and the return-on-equity factor play a role for several industries. Second, there seems to be quite some heterogeneity across different industries. While barely any variable is selected by Lasso for Utilities, many factors are selected for, e.g., High-Tech and Durable, but they do not coincide at all. In other words, there seems to be a clear picture that we do not need many factors, but Lasso does not provide a factor that consistently provides pricing abilities across industries.

14.4 Exercises

1. Write a function that requires three inputs, namely, `y` (a T vector), `X` (a $(T \times K)$ matrix), and `lambda` and then returns the Ridge estimator (a K vector) for a given penalization parameter λ. Recall that the intercept should not be penalized. Therefore, your function should indicate whether X contains a vector of ones as the first column, which should be exempt from the L_2 penalty.
2. Compute the L_2 norm ($\beta'\beta$) for the regression coefficients based on the predictive regression from the previous exercise for a range of λ's and illustrate the effect of penalization in a suitable figure.
3. Now, write a function that requires three inputs, namely,`y` (a T vector), `X` (a $(T \times K)$ matrix), and 'lambda' and then returns the Lasso estimator (a K vector) for a given penalization parameter λ. Recall that the intercept should not be penalized. Therefore, your function should indicate whether X contains a vector of ones as the first column, which should be exempt from the L_1 penalty.
4. After you understand what Ridge and Lasso regressions are doing, familiarize yourself with the `glmnet()` package's documentation. It is a thoroughly tested and well-established package that provides efficient code to compute the penalized regression coefficients for Ridge and Lasso and for combinations, commonly called *Elastic Nets*.

15

Option Pricing via Machine Learning

This chapter covers machine learning methods in option pricing. First, we briefly introduce regression trees, random forests, and neural networks – these methods are advocated as highly flexible *universal approximators*, capable of recovering highly nonlinear structures in the data. As the focus is on implementation, we leave a thorough treatment of the statistical underpinnings to other textbooks from authors with a real comparative advantage on these issues. We show how to implement random forests and deep neural networks with tidy principles using `tidymodels` or `TensorFlow` for more complicated network structures.

Machine learning (ML) is seen as a part of artificial intelligence. ML algorithms build a model based on training data in order to make predictions or decisions without being explicitly programmed to do so. While ML can be specified along a vast array of different branches, this chapter focuses on so-called supervised learning for regressions. The basic idea of supervised learning algorithms is to build a mathematical model for data that contains both the inputs and the desired outputs. In this chapter, we apply well-known methods such as random forests and neural networks to a simple application in option pricing. More specifically, we create an artificial dataset of option prices for different values based on the Black-Scholes pricing equation for call options. Then, we train different models to *learn* how to price call options without prior knowledge of the theoretical underpinnings of the famous option pricing equation by Black and Scholes (1973).

In order to replicate the analysis regarding neural networks in this chapter, you have to install `TensorFlow` on your system, which requires administrator rights on your machine. Parts of this can be done from within R. Just follow these quick-start instructions.[1]

Throughout this chapter, we need the following packages.

```
library(tidyverse)
library(tidymodels)
library(keras)
library(hardhat)
library(ranger)
```

[1] https://tensorflow.rstudio.com/installation/

DOI: 10.1201/b23237-15

The package keras (Allaire and Chollet, 2022) is a high-level neural networks API developed with a focus on enabling fast experimentation with Tensorflow. The package ranger (Wright and Ziegler, 2017) provides a fast implementation for random forests and hardhat (Vaughan and Kuhn, 2022) is a helper function to for robust data preprocessing at fit time and prediction time.

15.1 Regression Trees and Random Forests

Regression trees are a popular ML approach for incorporating multiway predictor interactions. In Finance, regression trees are gaining popularity, also in the context of asset pricing (see, e.g. Bryzgalova et al., 2022). Trees possess a logic that departs markedly from traditional regressions. Trees are designed to find groups of observations that behave similarly to each other. A tree *grows* in a sequence of steps. At each step, a new *branch* sorts the data leftover from the preceding step into bins based on one of the predictor variables. This sequential branching slices the space of predictors into partitions and approximates the unknown function $f(x)$ which yields the relation between the predictors x and the outcome variable y with the average value of the outcome variable within each partition. For a more thorough treatment of regression trees, we refer to Coqueret and Guida (2020).

Formally, we partition the predictor space into J non-overlapping regions, $R_1, R_2, \ldots, R_J$. For any predictor x that falls within region R_j, we estimate $f(x)$ with the average of the training observations, $\hat{y}_i$, for which the associated predictor x_i is also in R_j. Once we select a partition x to split in order to create the new partitions, we find a predictor j and value s that define two new partitions, called $R_1(j,s)$ and $R_2(j,s)$, which split our observations in the current partition by asking if x_j is bigger than s:

$$R_1(j,s) = \{x \mid x_j < s\} \text{ and } R_2(j,s) = \{x \mid x_j \geq s\}.$$

To pick j and s, we find the pair that minimizes the residual sum of square (RSS):

$$\sum_{i:\, x_i \in R_1(j,s)} (y_i - \hat{y}_{R_1})^2 + \sum_{i:\, x_i \in R_2(j,s)} (y_i - \hat{y}_{R_2})^2$$

As in Chapter 14 in the context of penalized regressions, the first relevant question is: what are the hyperparameter decisions? Instead of a regularization parameter, trees are fully determined by the number of branches used to generate a partition (sometimes one specifies the minimum number of observations in each final branch instead of the maximum number of branches).

Models with a single tree may suffer from high predictive variance. Random forests address these shortcomings of decision trees. The goal is to improve the predictive performance and reduce instability by averaging multiple decision trees. A forest basically implies creating many regression trees and averaging their predictions. To assure that the individual trees are not the same, we use a bootstrap to induce randomness. More specifically, we build B decision trees $T_1, \ldots, T_B$ using the training sample. For that purpose, we randomly select features to be included in the building of each tree. For each observation in the test set we then form a prediction $\hat{y} = \frac{1}{B}\sum_{i=1}^{B} \hat{y}_{T_i}$.

15.2 Neural Networks

Roughly speaking, neural networks propagate information from an input layer, through one or multiple hidden layers, to an output layer. While the number of units (neurons) in the input layer is equal to the dimension of the predictors, the output layer usually consists of one neuron (for regression) or multiple neurons for classification. The output layer predicts the future data, similar to the fitted value in a regression analysis. Neural networks have theoretical underpinnings as *universal approximators* for any smooth predictive association (Hornik, 1991). Their complexity, however, ranks neural networks among the least transparent, least interpretable, and most highly parameterized ML tools. In finance, applications of neural networks can be found in the context of many different contexts, e.g. Avramov et al. (2022a), Chen et al. (2019), and Gu et al. (2020).

Each neuron applies a nonlinear *activation function* f to its aggregated signal before sending its output to the next layer

$$x_k^l = f\left(\theta_0^k + \sum_{j=1}^{N^l} z_j \theta_{l,j}^k\right)$$

Here, θ are the parameters to fit, N^l denotes the number of units (a hyperparameter to tune), and z_j are the input variables which can be either the raw data or, in the case of multiple chained layers, the outcome from a previous layers $z_j = x_k - 1$. While the easiest case with $f(x) = \alpha + \beta x$ resembles linear regression, typical activation functions are sigmoid (i.e., $f(x) = (1 + e^{-x})^{-1}$) or ReLu (i.e., $f(x) = max(x, 0)$).

Neural networks gain their flexibility from chaining multiple layers together. Naturally, this imposes many degrees of freedom on the network architecture for which no clear theoretical guidance exists. The specification of a neural network requires, at a minimum, a stance on depth (number of hidden layers), the activation function, the number of neurons, the connection structure of the units (dense or sparse), and the application of regularization techniques to avoid overfitting. Finally, *learning* means to choose optimal parameters relying on numerical optimization, which often requires specifying an appropriate learning rate. Despite these computational challenges, implementation in R is not tedious at all because we can use the API to `TensorFlow`.

15.3 Option Pricing

To apply ML methods in a relevant field of finance, we focus on option pricing. The application in its core is taken from Hull (2020). In its most basic form, call options give the owner the right but not the obligation to buy a specific stock (the underlying) at a specific price (the strike price K) at a specific date (the exercise date T). The Black–Scholes price (Black and Scholes, 1973) of a call option for a non-dividend-paying underlying stock is given by

$$C(S,T) = \Phi(d_1)S - \Phi(d_1 - \sigma\sqrt{T})Ke^{-rT}$$
$$d_1 = \frac{1}{\sigma\sqrt{T}}\left(\ln\left(\frac{S}{K}\right) + \left(r_f + \frac{\sigma^2}{2}\right)T\right)$$

where $C(S,T)$ is the price of the option as a function of today's stock price of the underlying, S, with time to maturity T, r_f is the risk-free interest rate, and σ is the volatility of the underlying stock return. Φ is the cumulative distribution function of a standard normal random variable.

The Black-Scholes equation provides a way to compute the arbitrage-free price of a call option once the parameters S, K, r_f, T, and σ are specified (arguably, in a realistic context, all parameters are easy to specify except for σ which has to be estimated). A simple R function allows computing the price as we do below.

```
black_scholes_price <- function(S, K = 70, r = 0, T = 1, sigma = 0.2) {
  d1 <- (log(S / K) + (r + sigma^2 / 2) * T) / (sigma * sqrt(T))
  value <- S * pnorm(d1) - K * exp(-r * T) * pnorm(d1 - sigma * sqrt(T))

  return(value)
}
```

15.4 Learning Black-Scholes

We illustrate the concept of ML by showing how ML methods *learn* the Black-Scholes equation after observing some different specifications and corresponding prices without us revealing the exact pricing equation.

15.4.1 Data simulation

To that end, we start with simulated data. We compute option prices for call options for a grid of different combinations of times to maturity (T), risk-free rates (r), volatilities (sigma), strike prices (K), and current stock prices (S). In the code below, we add an idiosyncratic error term to each observation such that the prices considered do not exactly reflect the values implied by the Black-Scholes equation.

In order to keep the analysis reproducible, we use `set.seed()`. A random seed specifies the start point when a computer generates a random number sequence and ensures that our simulated data is the same across different machines.

```
set.seed(420)

option_prices <- expand_grid(
  S = 40:60,
  K = 20:90,
  r = seq(from = 0, to = 0.05, by = 0.01),
  T = seq(from = 3 / 12, to = 2, by = 1 / 12),
  sigma = seq(from = 0.1, to = 0.8, by = 0.1)
) |>
  mutate(
    black_scholes = black_scholes_price(S, K, r, T, sigma),
    observed_price = map(
      black_scholes,
      function(x) x + rnorm(2, sd = 0.15)
    )
  ) |>
  unnest(observed_price)
```

The code above generates more than 1.5 million random parameter constellations. For each of these values, two *observed* prices reflecting the Black-Scholes prices are given and a random innovation term *pollutes* the observed prices. The intuition of this application is simple: the simulated data provides many observations of option prices – by using the Black-Scholes equation we can evaluate the actual predictive performance of a ML method, which would be hard in a realistic context were the actual arbitrage-free price would be unknown.

Next, we split the data into a training set (which contains 1% of all the observed option prices) and a test set that will only be used for the final evaluation. Note that the entire grid of possible combinations contains 3148992 different specifications. Thus, the sample to learn the Black-Scholes price contains only 31,489 observations and is therefore relatively small.

```
split <- initial_split(option_prices, prop = 1 / 100)
```

We process the training dataset further before we fit the different ML models. We define a `recipe()` that defines all processing steps for that purpose. For our specific case, we want to explain the observed price by the five variables that enter the Black-Scholes equation. The *true* price (stored in column `black_scholes`) should obviously not be used to fit the model. The recipe also reflects that we standardize all predictors to ensure that each variable exhibits a sample average of zero and a sample standard deviation of one.

```
rec <- recipe(observed_price ~ .,
  data = option_prices
) |>
  step_rm(black_scholes) |>
  step_normalize(all_predictors())
```

15.4.2 Single layer networks and random forests

Next, we show how to fit a neural network to the data. Note that this requires that `keras` is installed on your local machine. The function `mlp()` from the package `parsnip` provides the functionality to initialize a single layer, feed-forward neural network. The specification below defines a single layer feed-forward neural network with 10 hidden units. We set the number of training iterations to `epochs = 500`. The option `set_mode("regression")` specifies a linear activation function for the output layer.

```
nnet_model <- mlp(
  epochs = 500,
  hidden_units = 10,
) |>
  set_mode("regression") |>
  set_engine("keras", verbose = FALSE)
```

The `verbose = FALSE` argument prevents logging the results to the console. We can follow the straightforward `tidymodel` workflow as in the chapter before: define a workflow, equip it with the recipe, and specify the associated model. Finally, fit the model with the training data.

```
nn_fit <- workflow() |>
  add_recipe(rec) |>
  add_model(nnet_model) |>
  fit(data = training(split))
```

Once you are familiar with the `tidymodels` workflow, it is a piece of cake to fit other models from the `parsnip` family. For instance, the model below initializes a random forest with 50 trees contained in the ensemble, where we require at least 2000 observations in a node. The random forests are trained using the package `ranger`, which is required to be installed in order to run the code below.

```
rf_model <- rand_forest(
  trees = 50,
  min_n = 2000
) |>
  set_engine("ranger") |>
  set_mode("regression")
```

Fitting the model follows exactly the same convention as for the neural network before.

```
rf_fit <- workflow() |>
  add_recipe(rec) |>
  add_model(rf_model) |>
  fit(data = training(split))
```

15.4.3 Deep neural networks

A deep neural network is a neural network with multiple layers between the input and output layers. By chaining multiple layers together, more complex structures can be represented with fewer parameters than simple shallow (one-layer) networks as the one implemented above. For instance, image or text recognition are typical tasks where deep neural networks are used (for applications of deep neural networks in finance, see, for instance, Jiang et al., 2022; Jensen et al., 2022a).

Note that while the `tidymodels` workflow is extremely convenient, these more sophisticated multi-layer (so-called *deep*) neural networks are not supported by `tidymodels` yet (as of September 2022). Instead, an implementation of a deep neural network in R requires additional computational tools. For that reason, the code snippet below illustrates how to initialize a sequential model with three hidden layers with 10 units per layer. The `keras` package provides a convenient interface and is flexible enough to handle different activation functions. The `compile()` command defines the loss function with which the model predictions are evaluated.

```
model <- keras_model_sequential() |>
  layer_dense(
    input_shape = 5,
    units = 10,
    activation = "sigmoid"
  ) |>
```

```
  layer_dense(units = 10, activation = "sigmoid") |>
  layer_dense(units = 10, activation = "sigmoid") |>
  layer_dense(units = 1, activation = "linear") |>
  compile(
    loss = "mean_squared_error"
  )
model
```

```
Model: "sequential_1"

______________________________________________________________________
 Layer (type)                  Output Shape                Param #
======================================================================
 dense_5 (Dense)               (None, 10)                  60
 dense_4 (Dense)               (None, 10)                  110
 dense_3 (Dense)               (None, 10)                  110
 dense_2 (Dense)               (None, 1)                   11
======================================================================
Total params: 291
Trainable params: 291
Non-trainable params: 0
______________________________________________________________________
```

To train the neural network, we provide the inputs (`x`) and the variable to predict (`y`) and then fit the parameters. Note the slightly tedious use of the method `extract_mold(nn_fit)`. Instead of simply using the *raw* data, we fit the neural network with the same processed data that is used for the single-layer feed-forward network. What is the difference to simply calling `x = training(data) |> select(-observed_price, -black_scholes)`? Recall that the recipe standardizes the variables such that all columns have unit standard deviation and zero mean. Further, it adds consistency if we ensure that all models are trained using the same recipe such that a change in the recipe is reflected in the performance of any model. A final note on a potentially irritating observation: `fit()` alters the `keras` model – this is one of the few instances, where a function in R alters the *input* such that after the function call the object `model` is not same anymore!

```
model |>
  fit(
    x = extract_mold(nn_fit)$predictors |> as.matrix(),
    y = extract_mold(nn_fit)$outcomes |> pull(observed_price),
    epochs = 500, verbose = FALSE
  )
```

15.4.4 Universal approximation

Before we evaluate the results, we implement one more model. In principle, any non-linear function can also be approximated by a linear model containing the input variables' polynomial expansions. To illustrate this, we first define a new recipe, `rec_linear`, which processes the training data even further. We include polynomials up to the fifth degree of each predictor and then add all possible pairwise interaction terms. The final recipe step, `step_lincomb()`, removes potentially redundant variables (for instance, the interaction between r^2 and r^3 is the same as the term r^5). We fit a Lasso regression model with a pre-specified penalty term (consult Chapter 14 on how to tune the model hyperparameters).

```
rec_linear <- rec |>
  step_poly(all_predictors(),
    degree = 5,
    options = list(raw = T)
  ) |>
  step_interact(terms = ~ all_predictors():all_predictors()) |>
  step_lincomb(all_predictors())

lm_model <- linear_reg(penalty = 0.01) |>
  set_engine("glmnet")

lm_fit <- workflow() |>
  add_recipe(rec_linear) |>
  add_model(lm_model) |>
  fit(data = training(split))
```

15.5 Prediction Evaluation

Finally, we collect all predictions to compare the *out-of-sample* prediction error evaluated on ten thousand new data points. Note that for the evaluation, we use the call to `extract_mold()` to ensure that we use the same pre-processing steps for the testing data across each model. We also use the somewhat advanced functionality in `forge()`, which provides an easy, consistent, and robust pre-processor at prediction time.

```
out_of_sample_data <- testing(split) |>
  slice_sample(n = 10000)

predictive_performance <- model |>
```

```
predict(forge(
    out_of_sample_data,
    extract_mold(nn_fit)$blueprint
  )$predictors |> as.matrix()) |>
  as.vector() |>
  tibble("Deep NN" = _) |>
  bind_cols(nn_fit |>
    predict(out_of_sample_data)) |>
  rename("Single layer" = .pred) |>
  bind_cols(lm_fit |> predict(out_of_sample_data)) |>
  rename("Lasso" = .pred) |>
  bind_cols(rf_fit |> predict(out_of_sample_data)) |>
  rename("Random forest" = .pred) |>
  bind_cols(out_of_sample_data) |>
  pivot_longer("Deep NN":"Random forest", names_to = "Model") |>
  mutate(
    moneyness = (S - K),
    pricing_error = abs(value - black_scholes)
  )
```

In the lines above, we use each of the fitted models to generate predictions for the entire test data set of option prices. We evaluate the absolute pricing error as one possible measure of pricing accuracy, defined as the absolute value of the difference between predicted option price and the theoretical correct option price from the Black-Scholes model. We show the results graphically in Figure 15.1.

```
predictive_performance |>
  ggplot(aes(
    x = moneyness,
    y = pricing_error,
    color = Model,
    linetype = Model
    )) +
  geom_jitter(alpha = 0.05) +
  geom_smooth(se = FALSE, method = "gam") +
  labs(
    x = "Moneyness (S - K)", color = NULL,
    y = "Absolut prediction error (USD)",
    title = "Prediction errors of call option prices for different models",
    linetype = NULL
  )
```

The results can be summarized as follows:

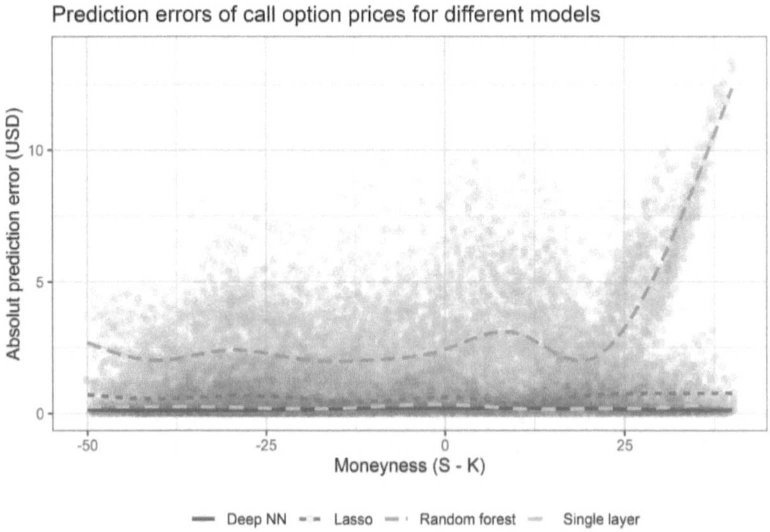

FIGURE 15.1
Absolut prediction error in USD for the different fitted methods. The prediction error is evaluated on a sample of call options that were not used for training.

1. All ML methods seem to be able to price call options after observing the training test set.
2. The average prediction errors increase for far in-the-money options.
3. Random forest and the Lasso seem to perform consistently worse in prediction option prices than the neural networks.
4. The complexity of the deep neural network relative to the single-layer neural network does not result in better out-of-sample predictions.

15.6 Exercises

1. Write a function that takes `y` and a matrix of predictors `X` as inputs and returns a characterization of the relevant parameters of a regression tree with 1 branch.
2. Create a function that creates predictions for a new matrix of predictors `newX` based on the estimated regression tree.
3. Use the package `rpart` to *grow* a tree based on the training data and use the illustration tools in `rpart` to understand which characteristics the tree deems relevant for option pricing.

4. Make use of a training and a test set to choose the optimal depth (number of sample splits) of the tree.
5. Use `keras` to initialize a sequential neural network that can take the predictors from the training data set as input, contains at least one hidden layer, and generates continuous predictions. *This sounds harder than it is:* see a simple regression example here.[2] How many parameters does the neural network you aim to fit have?
6. Compile the object from the previous exercise. It is important that you specify a loss function. Illustrate the difference in predictive accuracy for different architecture choices.

[2]https://tensorflow.rstudio.com/tutorials/beginners/basic-ml/tutorial_basic_regression/

Part V

Portfolio Optimization

16

Parametric Portfolio Policies

In this chapter, we apply different portfolio performance measures to evaluate and compare portfolio allocation strategies. For this purpose, we introduce a direct way to estimate optimal portfolio weights for large-scale cross-sectional applications. More precisely, the approach of Brandt et al. (2009) proposes to parametrize the optimal portfolio weights as a function of stock characteristics instead of estimating the stock's expected return, variance, and covariances with other stocks in a prior step. We choose weights as a function of the characteristics, which maximize the expected utility of the investor. This approach is feasible for large portfolio dimensions (such as the entire CRSP universe) and has been proposed by Brandt et al. (2009). See the review paper Brandt (2010) for an excellent treatment of related portfolio choice methods.

The current chapter relies on the following set of packages:

```
library(tidyverse)
library(lubridate)
library(RSQLite)
```

16.1 Data Preparation

To get started, we load the monthly CRSP file, which forms our investment universe. We load the data from our `SQLite`-database introduced in Chapters 2–4.

```
tidy_finance <- dbConnect(
  SQLite(), "data/tidy_finance.sqlite",
  extended_types = TRUE
)

crsp_monthly <- tbl(tidy_finance, "crsp_monthly") |>
  collect()
```

To evaluate the performance of portfolios, we further use monthly market returns as a benchmark to compute CAPM alphas.

DOI: 10.1201/b23237-16

```
factors_ff_monthly <- tbl(tidy_finance, "factors_ff_monthly") |>
  collect()
```

Next, we retrieve some stock characteristics that have been shown to have an effect on the expected returns or expected variances (or even higher moments) of the return distribution. In particular, we record the lagged one-year return momentum (`momentum_lag`), defined as the compounded return between months $t-12$ and $t-2$ for each firm. In finance, momentum is the empirically observed tendency for rising asset prices to rise further, and falling prices to keep falling (Jegadeesh and Titman, 1993). The second characteristic is the firm's market equity (`size_lag`), defined as the log of the price per share times the number of shares outstanding (Banz, 1981). To construct the correct lagged values, we use the approach introduced in Chapter 3.

```
crsp_monthly_lags <- crsp_monthly |>
  transmute(permno,
    month_12 = month %m+% months(12),
    mktcap
  )

crsp_monthly <- crsp_monthly |>
  inner_join(crsp_monthly_lags,
    by = c("permno", "month" = "month_12"),
    suffix = c("", "_12")
  )

data_portfolios <- crsp_monthly |>
  mutate(
    momentum_lag = mktcap_lag / mktcap_12,
    size_lag = log(mktcap_lag)
  ) |>
  drop_na(contains("lag"))
```

16.2 Parametric Portfolio Policies

The basic idea of parametric portfolio weights is as follows. Suppose that at each date t we have N_t stocks in the investment universe, where each stock i has a return of $r_{i,t+1}$ and is associated with a vector of firm characteristics $x_{i,t}$ such as time-series momentum or the market capitalization. The investor's problem is to choose portfolio weights $w_{i,t}$ to maximize the expected utility of the portfolio return:

$$\max_{w} E_t\left(u(r_{p,t+1})\right) = E_t\left[u\left(\sum_{i=1}^{N_t} w_{i,t} r_{i,t+1}\right)\right]$$

where $u(\cdot)$ denotes the utility function.

Where do the stock characteristics show up? We parameterize the optimal portfolio weights as a function of the stock characteristic $x_{i,t}$ with the following linear specification for the portfolio weights:

$$w_{i,t} = \bar{w}_{i,t} + \frac{1}{N_t}\theta' \hat{x}_{i,t},$$

where $\bar{w}_{i,t}$ is a stock's weight in a benchmark portfolio (we use the value-weighted or naive portfolio in the application below), θ is a vector of coefficients which we are going to estimate, and $\hat{x}_{i,t}$ are the characteristics of stock i, cross-sectionally standardized to have zero mean and unit standard deviation.

Intuitively, the portfolio strategy is a form of active portfolio management relative to a performance benchmark. Deviations from the benchmark portfolio are derived from the individual stock characteristics. Note that by construction the weights sum up to one as $\sum_{i=1}^{N_t} \hat{x}_{i,t} = 0$ due to the standardization. Moreover, the coefficients are constant across assets and over time. The implicit assumption is that the characteristics fully capture all aspects of the joint distribution of returns that are relevant for forming optimal portfolios.

We first implement cross-sectional standardization for the entire CRSP universe. We also keep track of (lagged) relative market capitalization `relative_mktcap`, which will represent the value-weighted benchmark portfolio, while `n` denotes the number of traded assets N_t, which we use to construct the naive portfolio benchmark.

```
data_portfolios <- data_portfolios |>
  group_by(month) |>
  mutate(
    n = n(),
    relative_mktcap = mktcap_lag / sum(mktcap_lag),
    across(contains("lag"), ~ (. - mean(.)) / sd(.)),
  ) |>
  ungroup() |>
  select(-mktcap_lag, -altprc)
```

16.3 Computing Portfolio Weights

Next, we move on to identify optimal choices of θ. We rewrite the optimization problem together with the weight parametrization and can then estimate θ to maximize the objective function based on our sample

$$E_t\left(u(r_{p,t+1})\right) = \frac{1}{T}\sum_{t=0}^{T-1} u\left(\sum_{i=1}^{N_t}\left(\bar{w}_{i,t} + \frac{1}{N_t}\theta'\hat{x}_{i,t}\right) r_{i,t+1}\right).$$

The allocation strategy is straightforward because the number of parameters to estimate is small. Instead of a tedious specification of the N_t dimensional vector of expected returns and the $N_t(N_t+1)/2$ free elements of the covariance matrix, all we need to focus on in our application is the vector θ. θ contains only two elements in our application – the relative deviation from the benchmark due to *size* and *momentum*.

To get a feeling for the performance of such an allocation strategy, we start with an arbitrary initial vector θ_0. The next step is to choose θ optimally to maximize the objective function. We automatically detect the number of parameters by counting the number of columns with lagged values.

```
n_parameters <- sum(str_detect(
  colnames(data_portfolios), "lag"
))

theta <- rep(1.5, n_parameters)

names(theta) <- colnames(data_portfolios)[str_detect(
  colnames(data_portfolios), "lag"
)]
```

The function `compute_portfolio_weights()` below computes the portfolio weights $\bar{w}_{i,t} + \frac{1}{N_t}\theta'\hat{x}_{i,t}$ according to our parametrization for a given value θ_0. Everything happens within a single pipeline. Hence, we provide a short walk-through.

We first compute `characteristic_tilt`, the tilting values $\frac{1}{N_t}\theta'\hat{x}_{i,t}$ which resemble the deviation from the benchmark portfolio. Next, we compute the benchmark portfolio `weight_benchmark`, which can be any reasonable set of portfolio weights. In our case, we choose either the value or equal-weighted allocation. `weight_tilt` completes the picture and contains the final portfolio weights `weight_tilt = weight_benchmark + characteristic_tilt` which deviate from the benchmark portfolio depending on the stock characteristics.

The final few lines go a bit further and implement a simple version of a no-short sale constraint. While it is generally not straightforward to ensure portfolio weight constraints via parameterization, we simply normalize the portfolio weights such that they are enforced to be positive. Finally, we make sure that the normalized weights sum up to one again:

$$w_{i,t}^{+} = \frac{\max(0, w_{i,t})}{\sum_{j=1}^{N_t} \max(0, w_{i,t})}.$$

The following function computes the optimal portfolio weights in the way just described.

```
compute_portfolio_weights <- function(theta,
                                      data,
                                      value_weighting = TRUE,
                                      allow_short_selling = TRUE) {
  data |>
    group_by(month) |>
    bind_cols(
      characteristic_tilt = data |>
        transmute(across(contains("lag"), ~ . / n)) |>
        as.matrix() %*% theta |> as.numeric()
    ) |>
    mutate(
      # Definition of benchmark weight
      weight_benchmark = case_when(
        value_weighting == TRUE ~ relative_mktcap,
        value_weighting == FALSE ~ 1 / n
      ),
      # Parametric portfolio weights
      weight_tilt = weight_benchmark + characteristic_tilt,
      # Short-sell constraint
      weight_tilt = case_when(
        allow_short_selling == TRUE ~ weight_tilt,
        allow_short_selling == FALSE ~ pmax(0, weight_tilt)
      ),
      # Weights sum up to 1
      weight_tilt = weight_tilt / sum(weight_tilt)
    ) |>
    ungroup()
}
```

In the next step, we compute the portfolio weights for the arbitrary vector θ_0. In the example below, we use the value-weighted portfolio as a benchmark and allow negative portfolio weights.

```
weights_crsp <- compute_portfolio_weights(theta,
  data_portfolios,
  value_weighting = TRUE,
  allow_short_selling = TRUE
)
```

16.4 Portfolio Performance

Are the computed weights optimal in any way? Most likely not, as we picked θ_0 arbitrarily. To evaluate the performance of an allocation strategy, one can think of many different approaches. In their original paper, Brandt et al. (2009) focus on a simple evaluation of the hypothetical utility of an agent equipped with a power utility function $u_\gamma(r) = \frac{(1+r)^\gamma}{1-\gamma}$, where γ is the risk aversion factor.

```
power_utility <- function(r, gamma = 5) {
  (1 + r)^(1 - gamma) / (1 - gamma)
}
```

We want to note that Gehrig et al. (2020) warn that, in the leading case of constant relative risk aversion (CRRA), strong assumptions on the properties of the returns, the variables used to implement the parametric portfolio policy, and the parameter space are necessary to obtain a well-defined optimization problem.

No doubt, there are many other ways to evaluate a portfolio. The function below provides a summary of all kinds of interesting measures that can be considered relevant. Do we need all these evaluation measures? It depends: the original paper Brandt et al. (2009) only cares about the expected utility to choose θ. However, if you want to choose optimal values that achieve the highest performance while putting some constraints on your portfolio weights, it is helpful to have everything in one function.

```
evaluate_portfolio <- function(weights_crsp,
                               full_evaluation = TRUE) {
  evaluation <- weights_crsp |>
    group_by(month) |>
    summarize(
      return_tilt = weighted.mean(ret_excess, weight_tilt),
      return_benchmark = weighted.mean(ret_excess, weight_benchmark)
    ) |>
    pivot_longer(-month,
      values_to = "portfolio_return",
```

```
    names_to = "model"
  ) |>
  group_by(model) |>
  left_join(factors_ff_monthly, by = "month") |>
  summarize(tibble(
    "Expected utility" = mean(power_utility(portfolio_return)),
    "Average return" = 100 * mean(12 * portfolio_return),
    "SD return" = 100 * sqrt(12) * sd(portfolio_return),
    "Sharpe ratio" = mean(portfolio_return) / sd(portfolio_return),
    "CAPM alpha" = coefficients(lm(portfolio_return ~ mkt_excess))[1],
    "Market beta" = coefficients(lm(portfolio_return ~ mkt_excess))[2]
  )) |>
  mutate(model = str_remove(model, "return_")) |>
  pivot_longer(-model, names_to = "measure") |>
  pivot_wider(names_from = model, values_from = value)

 if (full_evaluation) {
  weight_evaluation <- weights_crsp |>
    select(month, contains("weight")) |>
    pivot_longer(-month, values_to = "weight", names_to = "model") |>
    group_by(model, month) |>
    transmute(tibble(
      "Absolute weight" = abs(weight),
      "Max. weight" = max(weight),
      "Min. weight" = min(weight),
      "Avg. sum of negative weights" = -sum(weight[weight < 0]),
      "Avg. fraction of negative weights" = sum(weight < 0) / n()
    )) |>
    group_by(model) |>
    summarize(across(-month, ~ 100 * mean(.))) |>
    mutate(model = str_remove(model, "weight_")) |>
    pivot_longer(-model, names_to = "measure") |>
    pivot_wider(names_from = model, values_from = value)
  evaluation <- bind_rows(evaluation, weight_evaluation)
 }
 return(evaluation)
}
```

Let us take a look at the different portfolio strategies and evaluation measures.

```
evaluate_portfolio(weights_crsp) |>
  print(n = Inf)
```

```
# A tibble: 11 x 3
```

```
   measure                           benchmark     tilt
   <chr>                                 <dbl>    <dbl>
 1 Expected utility                   -0.249     -0.262
 2 Average return                      7.12      -0.445
 3 SD return                          15.3       21.0
 4 Sharpe ratio                        0.135     -0.00613
 5 CAPM alpha                          0.000123  -0.00582
 6 Market beta                         0.993      0.930
 7 Absolute weight                     0.0247     0.0632
 8 Max. weight                         3.54       3.67
 9 Min. weight                         0.0000277 -0.145
10 Avg. sum of negative weights        0         77.9
11 Avg. fraction of negative weights   0         49.4
```

The value-weighted portfolio delivers an annualized return of more than 6 percent and clearly outperforms the tilted portfolio, irrespective of whether we evaluate expected utility, the Sharpe ratio, or the CAPM alpha. We can conclude the market beta is close to one for both strategies (naturally almost identically 1 for the value-weighted benchmark portfolio). When it comes to the distribution of the portfolio weights, we see that the benchmark portfolio weight takes less extreme positions (lower average absolute weights and lower maximum weight). By definition, the value-weighted benchmark does not take any negative positions, while the tilted portfolio also takes short positions.

16.5 Optimal Parameter Choice

Next, we move to a choice of θ that actually aims to improve some (or all) of the performance measures. We first define a helper function `compute_objective_function()`, which we then pass to an optimizer.

```r
compute_objective_function <- function(theta,
                                       data,
                                       objective_measure = "Expected utility",
                                       value_weighting = TRUE,
                                       allow_short_selling = TRUE) {
  processed_data <- compute_portfolio_weights(
    theta,
    data,
    value_weighting,
    allow_short_selling
  )
```

```
  objective_function <- evaluate_portfolio(processed_data,
    full_evaluation = FALSE
  ) |>
    filter(measure == objective_measure) |>
    pull(tilt)

  return(-objective_function)
}
```

You may wonder why we return the negative value of the objective function. This is simply due to the common convention for optimization procedures to search for minima as a default. By minimizing the negative value of the objective function, we get the maximum value as a result. In its most basic form, R optimization relies on the function `optim()`. As main inputs, the function requires an initial guess of the parameters and the objective function to minimize. Now, we are fully equipped to compute the optimal values of $\hat{\theta}$, which maximize the hypothetical expected utility of the investor.

```
optimal_theta <- optim(
  par = theta,
  compute_objective_function,
  objective_measure = "Expected utility",
  data = data_portfolios,
  value_weighting = TRUE,
  allow_short_selling = TRUE
)

optimal_theta$par
```

```
momentum_lag     size_lag
      0.0713      -1.9487
```

The resulting values of $\hat{\theta}$ are easy to interpret: intuitively, expected utility increases by tilting weights from the value-weighted portfolio toward smaller stocks (negative coefficient for size) and toward past winners (positive value for momentum). Both findings are in line with the well-documented size effect (Banz, 1981) and the momentum anomaly (Jegadeesh and Titman, 1993).

16.6 More Model Specifications

How does the portfolio perform for different model specifications? For this purpose, we compute the performance of a number of different modeling choices based on the entire CRSP sample. The next code chunk performs all the heavy lifting.

```
full_model_grid <- expand_grid(
  value_weighting = c(TRUE, FALSE),
  allow_short_selling = c(TRUE, FALSE),
  data = list(data_portfolios)
) |>
  mutate(optimal_theta = pmap(
    .l = list(
      data,
      value_weighting,
      allow_short_selling
    ),
    .f = ~ optim(
      par = theta,
      compute_objective_function,
      data = ..1,
      objective_measure = "Expected utility",
      value_weighting = ..2,
      allow_short_selling = ..3
    )$par
  ))
```

Finally, we can compare the results. The table below shows summary statistics for all possible combinations: equal- or value-weighted benchmark portfolio, with or without short-selling constraints, and tilted toward maximizing expected utility.

```
performance_table <- full_model_grid |>
  mutate(
    processed_data = pmap(
      .l = list(
        optimal_theta,
        data,
        value_weighting,
        allow_short_selling
      ),
      .f = ~ compute_portfolio_weights(..1, ..2, ..3, ..4)
    ),
```

```
    portfolio_evaluation = map(processed_data,
      evaluate_portfolio,
      full_evaluation = TRUE
    )
  ) |>
  select(
    value_weighting,
    allow_short_selling,
    portfolio_evaluation
  ) |>
  unnest(portfolio_evaluation)

performance_table |>
  rename(
    " " = benchmark,
    Optimal = tilt
  ) |>
  mutate(
    value_weighting = case_when(
      value_weighting == TRUE ~ "VW",
      value_weighting == FALSE ~ "EW"
    ),
    allow_short_selling = case_when(
      allow_short_selling == TRUE ~ "",
      allow_short_selling == FALSE ~ "(no s.)"
    )
  ) |>
  pivot_wider(
    names_from = value_weighting:allow_short_selling,
    values_from = " ":Optimal,
    names_glue = "{value_weighting} {allow_short_selling} {.value} "
  ) |>
  select(
    measure,
    `EW    `,
    `VW    `,
    sort(contains("Optimal"))
  ) |>
  print(n = 11)
```

```
# A tibble: 11 x 7
  measure         `EW    ` `VW    ` VW  Op~1 VW (no~2 EW  Op~3 EW (no~4
  <chr>              <dbl>    <dbl>    <dbl>    <dbl>    <dbl>    <dbl>
1 Expected ut~ -0.250   -2.49e-1 -0.246   -0.247   -0.250   -2.50e-1
```

```
 2 Average ret~ 10.7       7.12e+0 14.9      13.6      13.1      8.14e+0
 3 SD return    20.3       1.53e+1 20.5      19.5      22.6      1.70e+1
 4 Sharpe ratio  0.152     1.35e-1  0.209     0.202     0.168    1.38e-1
 5 CAPM alpha    0.00226   1.23e-4  0.00646   0.00526   0.00428  4.69e-4
 6 Market beta   1.13      9.93e-1  1.01      1.04      1.14     1.08e+0
 7 Absolute we~  0.0247    2.47e-2  0.0373    0.0247    0.0255   2.47e-2
 8 Max. weight   0.0247    3.54e+0  3.37      2.69      0.0672   2.20e-1
 9 Min. weight   0.0247    2.77e-5 -0.0283    0        -0.0305   0
10 Avg. sum of~  0         0       26.4       0         1.73     0
11 Avg. fracti~  0         0       38.7       0         6.27     0
# ... with abbreviated variable names 1: `VW  Optimal `,
#   2: `VW (no s.) Optimal `, 3: `EW  Optimal `,
#   4: `EW (no s.) Optimal `
```

The results indicate that the average annualized Sharpe ratio of the equal-weighted portfolio exceeds the Sharpe ratio of the value-weighted benchmark portfolio. Nevertheless, starting with the weighted value portfolio as a benchmark and tilting optimally with respect to momentum and small stocks yields the highest Sharpe ratio across all specifications. Finally, imposing no short-sale constraints does not improve the performance of the portfolios in our application.

16.7 Exercises

1. How do the estimated parameters $\hat{\theta}$ and the portfolio performance change if your objective is to maximize the Sharpe ratio instead of the hypothetical expected utility?
2. The code above is very flexible in the sense that you can easily add new firm characteristics. Construct a new characteristic of your choice and evaluate the corresponding coefficient $\hat{\theta}_i$.
3. Tweak the function `optimal_theta()` such that you can impose additional performance constraints in order to determine $\hat{\theta}$, which maximizes expected utility under the constraint that the market beta is below 1.
4. Does the portfolio performance resemble a realistic out-of-sample backtesting procedure? Verify the robustness of the results by first estimating $\hat{\theta}$ based on *past data* only. Then, use more recent periods to evaluate the actual portfolio performance.
5. By formulating the portfolio problem as a statistical estimation problem, you can easily obtain standard errors for the coefficients of the weight function. Brandt et al. (2009) provide the relevant derivations in their paper in Equation (10). Implement a small function that computes standard errors for $\hat{\theta}$.

17

Constrained Optimization and Backtesting

In this chapter, we conduct portfolio backtesting in a realistic setting by including transaction costs and investment constraints such as no-short-selling rules. We start with standard mean-variance efficient portfolios and introduce constraints in a step-by-step manner. To do so, we rely on numerical optimization procedures in R. We conclude the chapter by providing an out-of-sample backtesting procedure for the different strategies that we introduce in this chapter.

Throughout this chapter, we use the following packages:

```
library(tidyverse)
library(RSQLite)
library(scales)
library(quadprog)
library(alabama)
```

Compared to previous chapters, we introduce the `quadprog` package (Turlach et al., 2019) to perform numerical constrained optimization for quadratic objective functions and `alabama` (Varadhan, 2022) for more general non-linear objective functions and constraints.

17.1 Data Preparation

We start by loading the required data from our `SQLite`-database introduced in Chapters 2–4. For simplicity, we restrict our investment universe to the monthly Fama-French industry portfolio returns in the following application.

```
tidy_finance <- dbConnect(
  SQLite(),
  "data/tidy_finance.sqlite",
  extended_types = TRUE
)
```

DOI: 10.1201/b23237-17

```
industry_returns <- tbl(tidy_finance, "industries_ff_monthly") |>
  collect()

industry_returns <- industry_returns |>
  select(-month)
```

17.2 Recap of Portfolio Choice

A common objective for portfolio optimization is to find mean-variance efficient portfolio weights, i.e., the allocation which delivers the lowest possible return variance for a given minimum level of expected returns. In the most extreme case, where the investor is only concerned about portfolio variance, she may choose to implement the minimum variance portfolio (MVP) weights which are given by the solution to

$$w_{\text{mvp}} = \arg\min w'\Sigma w \text{ s.t. } w'\iota = 1$$

where Σ is the $(N \times N)$ covariance matrix of the returns. The optimal weights ω_{mvp} can be found analytically and are $\omega_{\text{mvp}} = \frac{\Sigma^{-1}\iota}{\iota'\Sigma^{-1}\iota}$. In terms of code, the math is equivalent to the following chunk.

```
Sigma <- cov(industry_returns)
w_mvp <- solve(Sigma) %*% rep(1, ncol(Sigma))
w_mvp <- as.vector(w_mvp / sum(w_mvp))
```

Next, consider an investor who aims to achieve minimum variance *given a required expected portfolio return* $\bar{\mu}$ such that she chooses

$$w_{\text{eff}}(\bar{\mu}) = \arg\min w'\Sigma w \text{ s.t. } w'\iota = 1 \text{ and } \omega'\mu \geq \bar{\mu}.$$

We leave it as an exercise below to show that the portfolio choice problem can equivalently be formulated for an investor with mean-variance preferences and risk aversion factor γ. That means the investor aims to choose portfolio weights as the solution to

$$w^*_\gamma = \arg\max w'\mu - \frac{\gamma}{2}w'\Sigma w \quad \text{s.t. } w'\iota = 1.$$

The solution to the optimal portfolio choice problem is:

$$\omega^*_\gamma = \frac{1}{\gamma}\left(\Sigma^{-1} - \frac{1}{\iota'\Sigma^{-1}\iota}\Sigma^{-1}\iota\iota'\Sigma^{-1}\right)\mu + \frac{1}{\iota'\Sigma^{-1}\iota}\Sigma^{-1}\iota.$$

Empirically, this classical solution imposes many problems. In particular, the estimates of μ are noisy over short horizons, the $(N \times N)$ matrix Σ contains $N(N-1)/2$

distinct elements and thus, estimation error is huge. Seminal papers on the effect of ignoring estimation uncertainty, among others, are Brown (1976), Jobson and Korkie (1980), Jorion (1986), and Chopra and Ziemba (1993).

Even worse, if the asset universe contains more assets than available time periods ($N > T$), the sample covariance matrix is no longer positive definite such that the inverse Σ^{-1} does not exist anymore. To address estimation issues for vast-dimensional covariance matrices, regularization techniques are a popular tool (see, e.g., Ledoit and Wolf, 2003, 2004, 2012; Fan et al., 2008).

While the uncertainty associated with estimated parameters is challenging, the data-generating process is also unknown to the investor. In other words, model uncertainty reflects that it is ex-ante not even clear which parameters require estimation (for instance, if returns are driven by a factor model, selecting the universe of relevant factors imposes model uncertainty). Wang (2005) and Garlappi et al. (2007) provide theoretical analysis on optimal portfolio choice under model *and* estimation uncertainty. In the most extreme case, Pflug et al. (2012) shows that the naive portfolio which allocates equal wealth to all assets is the optimal choice for an investor averse to model uncertainty.

On top of the estimation uncertainty, *transaction costs* are a major concern. Rebalancing portfolios is costly, and, therefore, the optimal choice should depend on the investor's current holdings. In the presence of transaction costs, the benefits of reallocating wealth may be smaller than the costs associated with turnover. This aspect has been investigated theoretically, among others, for one risky asset by Magill and Constantinides (1976) and Davis and Norman (1990). Subsequent extensions to the case with multiple assets have been proposed by Balduzzi and Lynch (1999) and Balduzzi and Lynch (2000). More recent papers on empirical approaches which explicitly account for transaction costs include Gârleanu and Pedersen (2013), and DeMiguel et al. (2014), and DeMiguel et al. (2015).

17.3 Estimation Uncertainty and Transaction Costs

The empirical evidence regarding the performance of a mean-variance optimization procedure in which you simply plug in some sample estimates $\hat{\mu}$ and $\hat{\Sigma}$ can be summarized rather briefly: mean-variance optimization performs poorly! The literature discusses many proposals to overcome these empirical issues. For instance, one may impose some form of regularization of Σ, rely on Bayesian priors inspired by theoretical asset pricing models (Kan and Zhou, 2007) or use high-frequency data to improve forecasting (Hautsch et al., 2015). One unifying framework that works easily, effectively (even for large dimensions), and is purely inspired by economic arguments is an ex-ante adjustment for transaction costs (Hautsch and Voigt, 2019).

Assume that returns are from a multivariate normal distribution with mean μ and variance-covariance matrix Σ, $N(\mu, \Sigma)$. Additionally, we assume quadratic transaction costs which penalize rebalancing such that

$$\nu\left(\omega_{t+1}, \omega_{t^+}, \beta\right) = \frac{\beta}{2}\left(\omega_{t+1} - \omega_{t^+}\right)'\left(\omega_{t+1} - \omega_{t^+}\right),$$

with cost parameter $\beta > 0$ and $\omega_{t^+} = \omega_t \circ (1 + r_t)/\iota'(\omega_t \circ (1 + r_t))$. ω_{t^+} denotes the portfolio weights just before rebalancing. Note that ω_{t^+} differs mechanically from ω_t due to the returns in the past period. Intuitively, transaction costs penalize portfolio performance when the portfolio is shifted from the current holdings ω_{t^+} to a new allocation ω_{t+1}. In this setup, transaction costs do not increase linearly. Instead, larger rebalancing is penalized more heavily than small adjustments. Then, the optimal portfolio choice for an investor with mean variance preferences is

$$\begin{aligned}\omega_{t+1}^* &= \arg\max \omega'\mu - \nu_t(\omega, \omega_{t^+}, \beta) - \frac{\gamma}{2}\omega'\Sigma\omega \text{ s.t. } \iota'\omega = 1\\ &= \arg\max \omega'\mu^* - \frac{\gamma}{2}\omega'\Sigma^*\omega \text{ s.t.} \iota'\omega = 1,\end{aligned}$$

where

$$\mu^* = \mu + \beta\omega_{t^+} \quad \text{and} \quad \Sigma^* = \Sigma + \frac{\beta}{\gamma}I_N.$$

As a result, adjusting for transaction costs implies a standard mean-variance optimal portfolio choice with adjusted return parameters Σ^* and μ^*:

$$\omega_{t+1}^* = \frac{1}{\gamma}\left(\Sigma^{*-1} - \frac{1}{\iota'\Sigma^{*-1}\iota}\Sigma^{*-1}\iota\iota'\Sigma^{*-1}\right)\mu^* + \frac{1}{\iota'\Sigma^{*-1}\iota}\Sigma^{*-1}\iota.$$

An alternative formulation of the optimal portfolio can be derived as follows:

$$\omega_{t+1}^* = \arg\max \omega'\left(\mu + \beta\left(\omega_{t^+} - \frac{1}{N}\iota\right)\right) - \frac{\gamma}{2}\omega'\Sigma^*\omega \text{ s.t. } \iota'\omega = 1.$$

The optimal weights correspond to a mean-variance portfolio, where the vector of expected returns is such that assets that currently exhibit a higher weight are considered as delivering a higher expected return.

17.4 Optimal Portfolio Choice

The function below implements the efficient portfolio weight in its general form, allowing for transaction costs (conditional on the holdings *before* reallocation). For $\beta = 0$, the computation resembles the standard mean-variance efficient framework. `gamma` denotes the coefficient of risk aversion γ, `beta` is the transaction cost parameter β and `w_prev` are the weights before rebalancing ω_{t^+}.

```
compute_efficient_weight <- function(Sigma,
                                     mu,
                                     gamma = 2,
                                     beta = 0, # transaction costs
                                     w_prev = rep(
                                       1 / ncol(Sigma),
                                       ncol(Sigma)
                                     )) {
  iota <- rep(1, ncol(Sigma))
  Sigma_processed <- Sigma + beta / gamma * diag(ncol(Sigma))
  mu_processed <- mu + beta * w_prev

  Sigma_inverse <- solve(Sigma_processed)

  w_mvp <- Sigma_inverse %*% iota
  w_mvp <- as.vector(w_mvp / sum(w_mvp))
  w_opt <- w_mvp + 1 / gamma *
    (Sigma_inverse - w_mvp %*% t(iota) %*% Sigma_inverse) %*%
      mu_processed
  return(as.vector(w_opt))
}

mu <- colMeans(industry_returns)
compute_efficient_weight(Sigma, mu)
```

```
 [1]  1.395  0.293 -1.390  0.477  0.363 -0.320  0.545  0.446 -0.132
[10] -0.675
```

The portfolio weights above indicate the efficient portfolio for an investor with risk aversion coefficient $\gamma = 2$ in absence of transaction costs. Some of the positions are negative which implies short-selling, most of the positions are rather extreme. For instance, a position of -1 implies that the investor takes a short position worth her entire wealth to lever long positions in other assets. What is the effect of transaction costs or different levels of risk aversion on the optimal portfolio choice? The following few lines of code analyze the distance between the minimum variance portfolio and the portfolio implemented by the investor for different values of the transaction cost parameter β and risk aversion γ.

```
transaction_costs <- expand_grid(
  gamma = c(2, 4, 8, 20),
  beta = 20 * qexp((1:99) / 100)
) |>
  mutate(
    weights = map2(
```

```
    .x = gamma,
    .y = beta,
    ~ compute_efficient_weight(Sigma,
      mu,
      gamma = .x,
      beta = .y / 10000,
      w_prev = w_mvp
    )
  ),
  concentration = map_dbl(weights, ~ sum(abs(. - w_mvp)))
)
```

The code chunk above computes the optimal weight in presence of transaction cost for different values of β and γ but with the same initial allocation, the theoretical optimal minimum variance portfolio. Starting from the initial allocation, the investor chooses her optimal allocation along the efficient frontier to reflect her own risk preferences. If transaction costs would be absent, the investor would simply implement the mean-variance efficient allocation. If transaction costs make it costly to rebalance, her optimal portfolio choice reflects a shift toward the efficient portfolio, whereas her current portfolio anchors her investment.

```
transaction_costs |>
  mutate(risk_aversion = as_factor(gamma)) |>
  ggplot(aes(
    x = beta,
    y = concentration,
    color = risk_aversion,
    linetype = risk_aversion
  )) +
  geom_line() +
  guides(linetype = "none") +
  labs(
    x = "Transaction cost parameter",
    y = "Distance from MVP",
    color = "Risk aversion",
    title = "Portfolio weights for different risk aversion and transaction cost"
  )
```

Figure 17.1 shows rebalancing from the initial portfolio (which we always set to the minimum variance portfolio weights in this example). The higher the transaction costs parameter β, the smaller is the rebalancing from the initial portfolio. In addition, if risk aversion γ increases, the efficient portfolio is closer to the minimum variance portfolio weights such that the investor desires less rebalancing from the initial holdings.

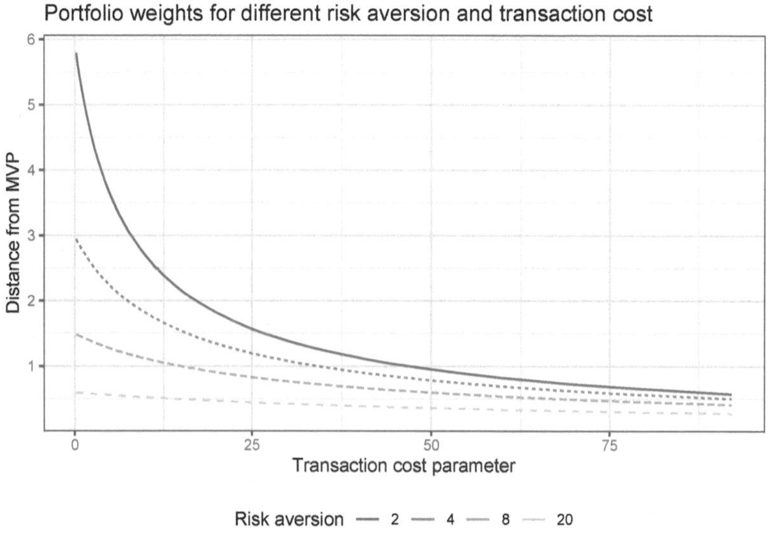

FIGURE 17.1
The horizontal axis indicates the distance from the empirical minimum variance portfolio weight, measured by the sum of the absolute deviations of the chosen portfolio from the benchmark.

17.5 Constrained Optimization

Next, we introduce constraints to the above optimization procedure. Very often, typical constraints such as short-selling restrictions prevent analytical solutions for optimal portfolio weights (short-selling restrictions simply imply that negative weights are not allowed such that we require that $w_i \geq 0 \forall i$). However, numerical optimization allows computing the solutions to such constrained problems. For the purpose of mean-variance optimization, we rely on the `solve.QP()` function from the package `quadprog`.

The function `solve.QP()` delivers numerical solutions to quadratic programming problems of the form

$$\min(-\mu\omega + 1/2\omega'\Sigma\omega) \text{ s.t. } A'\omega >= b_0.$$

The function takes one argument (`meq`) for the number of equality constraints. Therefore, the above matrix A is simply a vector of ones to ensure that the weights sum up to one. In the case of short-selling constraints, the matrix A is of the form

$$A' = \begin{pmatrix} 1 & 1 & \dots & 1 \\ 1 & 0 & \dots & 0 \\ 0 & 1 & \dots & 0 \\ \vdots & & \ddots & \vdots \\ 0 & 0 & \dots & 1 \end{pmatrix}' \qquad b_0 = \begin{pmatrix} 1 \\ 0 \\ \vdots \\ 0 \end{pmatrix}.$$

Before we dive into constrained optimization, we revisit the *unconstrained* problem and replicate the analytical solutions for the minimum variance and efficient portfolio weights from above. We verify that the output is equal to the above solution. Note that `near()` is a safe way to compare two vectors for pairwise equality. The alternative `==` is sensitive to small differences that may occur due to the representation of floating points on a computer, while `near()` has a built-in tolerance. As just discussed, we set `Amat` to a matrix with a column of ones and `bvec` to 1 to enforce the constraint that weights must sum up to one. `meq=1` means that one (out of one) constraints must be satisfied with equality.

```
n_industries <- ncol(industry_returns)

w_mvp_numerical <- solve.QP(
  Dmat = Sigma,
  dvec = rep(0, n_industries),
  Amat = cbind(rep(1, n_industries)),
  bvec = 1,
  meq = 1
)

all(near(w_mvp, w_mvp_numerical$solution))
```

```
[1] TRUE
```

```
w_efficient_numerical <- solve.QP(
  Dmat = 2 * Sigma,
  dvec = mu,
  Amat = cbind(rep(1, n_industries)),
  bvec = 1,
  meq = 1
)

all(near(compute_efficient_weight(Sigma, mu), w_efficient_numerical$solution))
```

```
[1] TRUE
```

The result above shows that indeed the numerical procedure recovered the optimal weights for a scenario, where we already know the analytic solution. For more complex optimization routines, R's optimization task view[1] provides an overview of the vast optimization landscape.

Next, we approach problems where no analytical solutions exist. First, we additionally impose short-sale constraints, which implies N inequality constraints of the form $w_i >= 0$.

```
w_no_short_sale <- solve.QP(
  Dmat = 2 * Sigma,
  dvec = mu,
  Amat = cbind(1, diag(n_industries)),
  bvec = c(1, rep(0, n_industries)),
  meq = 1
)
w_no_short_sale$solution
```

```
[1] 5.17e-01 3.06e-18 4.27e-16 7.90e-02 3.47e-18 2.65e-17 1.48e-01
[8] 2.56e-01 8.74e-18 0.00e+00
```

As expected, the resulting portfolio weights are all positive (up to numerical precision). Typically, the holdings in the presence of short-sale constraints are concentrated among way fewer assets than for the unrestricted case. You can verify that `sum(w_no_short_sale$solution)` returns 1. In other words: `solve.QP()` provides the numerical solution to a portfolio choice problem for a mean-variance investor with risk aversion `gamma = 2`, where negative holdings are forbidden.

`solve.QP()` is fast because it benefits from a very clear problem structure with a quadratic objective and linear constraints. However, optimization often requires more flexibility. As an example, we show how to compute optimal weights, subject to the so-called Regulation T-constraint,[2] which requires that the sum of all absolute portfolio weights is smaller than 1.5, that is $\sum_{i=1}^{N} |w_i| \leq 1.5$. The constraint enforces that a maximum of 50 percent of the allocated wealth can be allocated to short positions, thus implying an initial margin requirement of 50 percent. Imposing such a margin requirement reduces portfolio risks because extreme portfolio weights are not attainable anymore. The implementation of Regulation-T rules is numerically interesting because the margin constraints imply a non-linear constraint on the portfolio weights. Thus, we can no longer rely on `solve.QP()`, which is defined as solving quadratic programming problems with linear constraints. Instead, we rely on the package `alabama`, which requires a separate definition of objective and constraint functions.

[1] https://cran.r-project.org/web/views/Optimization.html
[2] https://en.wikipedia.org/wiki/Regulation_T

```
initial_weights <- rep(
  1 / n_industries,
  n_industries
)

objective <- function(w, gamma = 2) {
  -t(w) %*% (1 + mu) +
    gamma / 2 * t(w) %*% Sigma %*% w
}

inequality_constraints <- function(w, reg_t = 1.5) {
  reg_t - sum(abs(w))
}

equality_constraints <- function(w) {
  sum(w) - 1
}

w_reg_t <- constrOptim.nl(
  par = initial_weights,
  hin = inequality_constraints,
  fn = objective,
  heq = equality_constraints,
  control.outer = list(trace = FALSE)
)
w_reg_t$par
```

```
 [1]  3.41e-01 -1.81e-06 -1.08e-01  1.40e-01  7.30e-02 -1.08e-02
 [7]  2.46e-01  3.14e-01  1.35e-01 -1.30e-01
```

Note that the function `constrOptim.nl()` requires a starting vector of parameter values, i.e., an initial portfolio. Under the hood, `alamaba` performs numerical optimization by searching for a local minimum of the function `objective()` (subject to the equality constraints `equality_constraints()` and the inequality constraints `inequality_constraints()`). Note that the starting point should not matter if the algorithm identifies a global minimum.

Figure 17.2 shows the optimal allocation weights across all 10 industries for the four different strategies considered so far: minimum variance, efficient portfolio with $\gamma = 2$, efficient portfolio with short-sale constraints, and the Regulation-T constrained portfolio.

```
tibble(
  `No short-sale` = w_no_short_sale$solution,
  `Minimum Variance` = w_mvp,
```

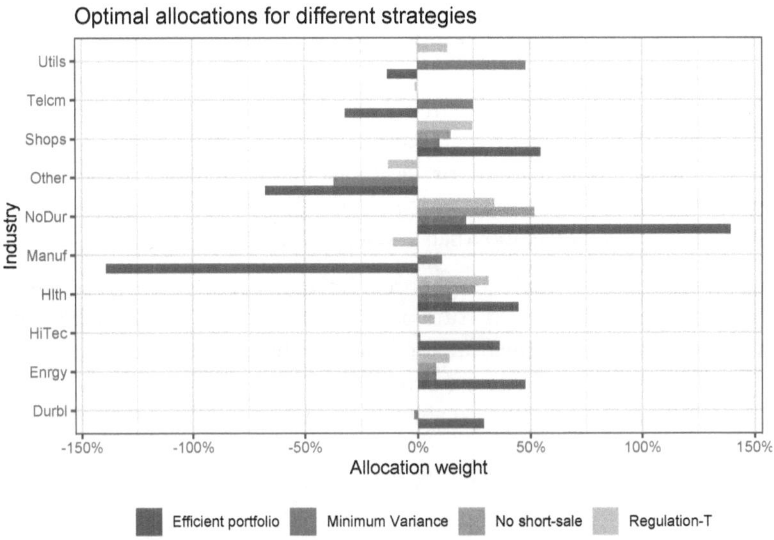

FIGURE 17.2
Optimal allocation weights for the 10 industry portfolios and the 4 different allocation strategies.

```
  `Efficient portfolio` = compute_efficient_weight(Sigma, mu),
  `Regulation-T` = w_reg_t$par,
  Industry = colnames(industry_returns)
) |>
  pivot_longer(-Industry,
    names_to = "Strategy",
    values_to = "weights"
  ) |>
  ggplot(aes(
    fill = Strategy,
    y = weights,
    x = Industry
  )) +
  geom_bar(position = "dodge", stat = "identity") +
  coord_flip() +
  labs(
    y = "Allocation weight", fill = NULL,
    title = "Optimal allocations for different strategies"
  ) +
  scale_y_continuous(labels = percent)
```

The results clearly indicate the effect of imposing additional constraints: the extreme holdings the investor implements if she follows the (theoretically optimal) efficient portfolio vanish under, e.g., the Regulation-T constraint. You may wonder why an investor would deviate from what is theoretically the optimal portfolio by imposing potentially arbitrary constraints. The short answer is: the *efficient portfolio* is only efficient if the true parameters of the data generating process correspond to the estimated parameters $\hat{\Sigma}$ and $\hat{\mu}$. Estimation uncertainty may thus lead to inefficient allocations. By imposing restrictions, we implicitly shrink the set of possible weights and prevent extreme allocations, which could result from *error-maximization* due to estimation uncertainty (Jagannathan and Ma, 2003).

Before we move on, we want to propose a final allocation strategy, which reflects a somewhat more realistic structure of transaction costs instead of the quadratic specification used above. The function below computes efficient portfolio weights while adjusting for transaction costs of the form $\beta\sum_{i=1}^{N}|(w_{i,t+1}-w_{i,t^+})|$. No closed-form solution exists, and we rely on non-linear optimization procedures.

```
compute_efficient_weight_L1_TC <- function(mu,
                                           Sigma,
                                           gamma = 2,
                                           beta = 0,
                                           initial_weights = rep(
                                             1 / ncol(Sigma),
                                             ncol(Sigma)
                                           )) {
  objective <- function(w) {
    -t(w) %*% mu +
      gamma / 2 * t(w) %*% Sigma %*% w +
      (beta / 10000) / 2 * sum(abs(w - initial_weights))
  }

  w_optimal <- constrOptim.nl(
    par = initial_weights,
    fn = objective,
    heq = function(w) {
      sum(w) - 1
    },
    control.outer = list(trace = FALSE)
  )

  return(w_optimal$par)
}
```

17.6 Out-of-Sample Backtesting

For the sake of simplicity, we committed one fundamental error in computing portfolio weights above: we used the full sample of the data to determine the optimal allocation (Arnott et al., 2019). To implement this strategy at the beginning of the 2000s, you will need to know how the returns will evolve until 2021. While interesting from a methodological point of view, we cannot evaluate the performance of the portfolios in a reasonable out-of-sample fashion. We do so next in a backtesting application for three strategies. For the backtest, we recompute optimal weights just based on past available data.

The few lines below define the general setup. We consider 120 periods from the past to update the parameter estimates before recomputing portfolio weights. Then, we update portfolio weights which is costly and affects the performance. The portfolio weights determine the portfolio return. A period later, the current portfolio weights have changed and form the foundation for transaction costs incurred in the next period. We consider three different competing strategies: the mean-variance efficient portfolio, the mean-variance efficient portfolio with ex-ante adjustment for transaction costs, and the naive portfolio, which allocates wealth equally across the different assets.

```
window_length <- 120
periods <- nrow(industry_returns) - window_length

beta <- 50
gamma <- 2

performance_values <- matrix(NA,
  nrow = periods,
  ncol = 3
)
colnames(performance_values) <- c("raw_return", "turnover", "net_return")

performance_values <- list(
  "MV (TC)" = performance_values,
  "Naive" = performance_values,
  "MV" = performance_values
)

w_prev_1 <- w_prev_2 <- w_prev_3 <- rep(
  1 / n_industries,
  n_industries
)
```

We also define two helper functions: one to adjust the weights due to returns and one for performance evaluation, where we compute realized returns net of transaction costs.

```
adjust_weights <- function(w, next_return) {
  w_prev <- 1 + w * next_return
  as.numeric(w_prev / sum(as.vector(w_prev)))
}

evaluate_performance <- function(w, w_previous, next_return, beta = 50) {
  raw_return <- as.matrix(next_return) %*% w
  turnover <- sum(abs(w - w_previous))
  net_return <- raw_return - beta / 10000 * turnover
  c(raw_return, turnover, net_return)
}
```

The following code chunk performs a rolling-window estimation, which we implement in a loop. In each period, the estimation window contains the returns available up to the current period. Note that we use the sample variance-covariance matrix and ignore the estimation of $\hat{\mu}$ entirely, but you might use more advanced estimators in practice.

```
for (p in 1:periods) {
  returns_window <- industry_returns[p:(p + window_length - 1), ]
  next_return <- industry_returns[p + window_length, ] |> as.matrix()

  Sigma <- cov(returns_window)
  mu <- 0 * colMeans(returns_window)

  # Transaction-cost adjusted portfolio
  w_1 <- compute_efficient_weight_L1_TC(
    mu = mu,
    Sigma = Sigma,
    beta = beta,
    gamma = gamma,
    initial_weights = w_prev_1
  )

  performance_values[[1]][p, ] <- evaluate_performance(w_1,
    w_prev_1,
    next_return,
    beta = beta
  )

  w_prev_1 <- adjust_weights(w_1, next_return)
```

```
  # Naive portfolio
  w_2 <- rep(1 / n_industries, n_industries)

  performance_values[[2]][p, ] <- evaluate_performance(
    w_2,
    w_prev_2,
    next_return
  )

  w_prev_2 <- adjust_weights(w_2, next_return)

  # Mean-variance efficient portfolio (w/o transaction costs)
  w_3 <- compute_efficient_weight(
    Sigma = Sigma,
    mu = mu,
    gamma = gamma
  )

  performance_values[[3]][p, ] <- evaluate_performance(
    w_3,
    w_prev_3,
    next_return
  )

  w_prev_3 <- adjust_weights(w_3, next_return)
}
```

Finally, we get to the evaluation of the portfolio strategies *net-of-transaction costs*. Note that we compute annualized returns and standard deviations.

```
performance <- lapply(
  performance_values,
  as_tibble
) |>
  bind_rows(.id = "strategy")

performance |>
  group_by(strategy) |>
  summarize(
    Mean = 12 * mean(100 * net_return),
    SD = sqrt(12) * sd(100 * net_return),
    `Sharpe ratio` = if_else(Mean > 0,
      Mean / SD,
      NA_real_
```

```
  ),
  Turnover = 100 * mean(turnover)
)
```

```
# A tibble: 3 x 5
  strategy    Mean    SD `Sharpe ratio` Turnover
  <chr>      <dbl> <dbl>          <dbl>    <dbl>
1 MV        -0.635  12.5         NA     213.
2 MV (TC)   12.3    15.0          0.820   0.0298
3 Naive     12.3    15.0          0.818   0.230
```

The results clearly speak against mean-variance optimization. Turnover is huge when the investor only considers her portfolio's expected return and variance. Effectively, the mean-variance portfolio generates a *negative* annualized return after adjusting for transaction costs. At the same time, the naive portfolio turns out to perform very well. In fact, the performance gains of the transaction-cost adjusted mean-variance portfolio are small. The out-of-sample Sharpe ratio is slightly higher than for the naive portfolio. Note the extreme effect of turnover penalization on turnover: *MV (TC)* effectively resembles a buy-and-hold strategy which only updates the portfolio once the estimated parameters $\hat{\mu}_t$ and $\hat{\Sigma}_t$indicate that the current allocation is too far away from the optimal theoretical portfolio.

17.7 Exercises

1. We argue that an investor with a quadratic utility function with certainty equivalent

$$\max_w CE(w) = \omega'\mu - \frac{\gamma}{2}\omega'\Sigma\omega \text{ s.t. } \iota'\omega = 1$$

faces an equivalent optimization problem to a framework, where portfolio weights are chosen with the aim to minimize volatility given a prespecified level or expected returns

$$\min_w \omega'\Sigma\omega \text{ s.t. } \omega'\mu = \bar{\mu} \text{ and } \iota'\omega = 1.$$

Proof that there is an equivalence between the optimal portfolio weights in both cases.

2. Consider the portfolio choice problem for transaction-cost adjusted certainty equivalent maximization with risk aversion parameter γ

$$\omega^*_{t+1} = \arg\max_{\omega\in\mathbb{R}^N, \iota'\omega=1} \omega'\mu - \nu_t(\omega, \beta) - \frac{\gamma}{2}\omega'\Sigma\omega$$

where Σ and μ are (estimators of) the variance-covariance matrix of the returns and the vector of expected returns. Assume for now that transaction costs are quadratic in rebalancing *and* proportional to stock illiquidity such that

$$\nu_t\left(\omega, B\right) = \frac{\beta}{2}\left(\omega - \omega_{t+}\right)' B \left(\omega - \omega_{t+}\right)$$

where $B = \text{diag}(ill_1, \ldots, ill_N)$ is a diagonal matrix, where $ill_1, \ldots, ill_N$. Derive a closed-form solution for the mean-variance efficient portfolio ω^*_{t+1} based on the transaction cost specification above. Discuss the effect of illiquidity ill_i on the individual portfolio weights relative to an investor that myopically ignores transaction costs in her decision.

3. Use the solution from the previous exercise to update the function `compute_efficient_weight()` such that you can compute optimal weights conditional on a matrix B with illiquidity measures.
4. Illustrate the evolution of the *optimal* weights from the naive portfolio to the efficient portfolio in the mean-standard deviation diagram.
5. Is it always optimal to choose the same β in the optimization problem than the value used in evaluating the portfolio performance? In other words: can it be optimal to choose theoretically sub-optimal portfolios based on transaction cost considerations that do not reflect the actual incurred costs? Evaluate the out-of-sample Sharpe ratio after transaction costs for a range of different values of imposed β values.

A

Cover Design

The cover of the book is inspired by the fast growing generative art community in R. Generative art refers to art that in whole or in part has been created with the use of an autonomous system. Instead of creating random dynamics we rely on what is core to the book: The evolution of financial markets. Each circle in the cover figure corresponds to daily market return within one year of our sample. Deviations from the circle line indicate positive or negative returns. The colors are determined by the standard deviation of market returns during the particular year. The few lines of code below replicate the entire figure. We use the Wes Andersen color palette (also throughout the entire book), provided by the package ´wesanderson´ (Ram and Wickham, 2018)

```
library(tidyverse)
library(lubridate)
library(RSQLite)
library(wesanderson)

tidy_finance <- dbConnect(
  SQLite(),
  "data/tidy_finance.sqlite",
  extended_types = TRUE
)

factors_ff_daily <- tbl(
  tidy_finance,
  "factors_ff_daily"
) |>
  collect()

data_plot <- factors_ff_daily |>
  select(date, mkt_excess) |>
  group_by(year = floor_date(date, "year")) |>
  mutate(group_id = cur_group_id())

data_plot <- data_plot |>
  group_by(group_id) |>
  mutate(
```

DOI: 10.1201/b23237-A

```
    day = 2 * pi * (1:n()) / 252,
    ymin = pmin(1 + mkt_excess, 1),
    ymax = pmax(1 + mkt_excess, 1),
    vola = sd(mkt_excess)
  ) |>
  filter(year >= "1962-01-01")

levels <- data_plot |>
  distinct(group_id, vola) |>
  arrange(vola) |>
  pull(vola)

cp <- coord_polar(
  direction = -1,
  clip = "on"
)

cp$is_free <- function() TRUE

colors <- wes_palette("Zissou1",
  n_groups(data_plot),
  type = "continuous"
)

plot <- data_plot |>
  mutate(vola = factor(vola, levels = levels)) |>
  ggplot(aes(
    x = day,
    y = mkt_excess,
    group = group_id,
    fill = vola
  )) +
  cp +
  geom_ribbon(aes(
    ymin = ymin,
    ymax = ymax,
    fill = vola
  ), alpha = 0.90) +
  theme_void() +
  facet_wrap(~group_id,
    ncol = 10,
    scales = "free"
  ) +
  theme(
```

```
    strip.text.x = element_blank(),
    legend.position = "None",
    panel.spacing = unit(-5, "lines")
  ) +
  scale_fill_manual(values = colors)

ggsave(
 plot = plot,
 width = 10,
 height = 6,
 filename = "cover.jpg",
 bg = "white"
)
```

B

Clean Enhanced TRACE with R

This appendix contains code to clean enhanced TRACE with R. It is also available via the following Github gist[1]. Hence, you could also source the function with `devtools::source_gist("3a05b3ab281563b2e94858451c2eb3a4")`. We need this function in Chapter 4 to download and clean enhanced TRACE trade messages following Dick-Nielsen (2009) and Dick-Nielsen (2014) for enhanced TRACE specifically. WRDS provides SAS code to clean enhanced TRACE data.

The function takes a vector of CUSIPs (in `cusips`), a connection to WRDS (`connection`) explained in Chapter 3, and a start and end date (`start_date` and `end_date`, respectively). Specifying too many CUSIPs will result in very slow downloads and a potential failure due to the size of the request to WRDS. The dates should be within the coverage of TRACE itself, i.e., starting after 2002, and the dates should be supplied using the class date. The output of the function contains all valid trade messages for the selected CUSIPs over the specified period.

```
clean_enhanced_trace <- function(cusips,
                                 connection,
                                 start_date = as.Date("2002-01-01"),
                                 end_date = today()) {


  # Packages (required)
  library(tidyverse)
  library(lubridate)
  library(dbplyr)
  library(RPostgres)

  # Function checks ---------------------------------------------------------
  # Input parameters
  ## Cusips
  if (length(cusips) == 0 | any(is.na(cusips))) stop("Check cusips.")

  ## Dates
  if (!is.Date(start_date) | !is.Date(end_date)) stop("Dates needed")
  if (start_date < as.Date("2002-01-01")) stop("TRACE starts later.")
```

[1] https://gist.github.com/patrick-weiss/3a05b3ab281563b2e94858451c2eb3a4

DOI: 10.1201/b23237-B

```r
if (end_date > today()) stop("TRACE does not predict the future.")
if (start_date >= end_date) stop("Date conflict.")

## Connection
if (!dbIsValid(connection)) stop("Connection issue.")

# Enhanced Trace ------------------------------------------------------------
# Main file
trace_all <- tbl(
  connection,
  in_schema("trace", "trace_enhanced")
) |>
  filter(cusip_id %in% cusips) |>
  filter(trd_exctn_dt >= start_date & trd_exctn_dt <= end_date) |>
  select(
    cusip_id, msg_seq_nb, orig_msg_seq_nb,
    entrd_vol_qt, rptd_pr, yld_pt, rpt_side_cd, cntra_mp_id,
    trd_exctn_dt, trd_exctn_tm, trd_rpt_dt, trd_rpt_tm,
    pr_trd_dt, trc_st, asof_cd, wis_fl,
    days_to_sttl_ct, stlmnt_dt, spcl_trd_fl
  ) |>
  collect()

# Enhanced Trace: Post 06-02-2012 -----------------------------------------
# Trades (trc_st = T) and correction (trc_st = R)
trace_post_TR <- trace_all |>
  filter(
    (trc_st == "T" | trc_st == "R"),
    trd_rpt_dt >= as.Date("2012-02-06")
  )

# Cancelations (trc_st = X) and correction cancelations (trc_st = C)
trace_post_XC <- trace_all |>
  filter(
    (trc_st == "X" | trc_st == "C"),
    trd_rpt_dt >= as.Date("2012-02-06")
  )

# Cleaning corrected and cancelled trades
trace_post_TR <- trace_post_TR |>
  anti_join(trace_post_XC,
    by = c(
      "cusip_id", "msg_seq_nb", "entrd_vol_qt",
      "rptd_pr", "rpt_side_cd", "cntra_mp_id",
```

```
      "trd_exctn_dt", "trd_exctn_tm"
    )
  )

# Reversals (trc_st = Y)
trace_post_Y <- trace_all |>
  filter(
    trc_st == "Y",
    trd_rpt_dt >= as.Date("2012-02-06")
  )

# Clean reversals
## match the orig_msg_seq_nb of the Y-message to
## the msg_seq_nb of the main message
trace_post <- trace_post_TR |>
  anti_join(trace_post_Y,
    by = c("cusip_id",
      "msg_seq_nb" = "orig_msg_seq_nb",
      "entrd_vol_qt", "rptd_pr", "rpt_side_cd",
      "cntra_mp_id", "trd_exctn_dt", "trd_exctn_tm"
    )
  )


# Enhanced TRACE: Pre 06-02-2012 -------------------------------------------
# Cancelations (trc_st = C)
trace_pre_C <- trace_all |>
  filter(
    trc_st == "C",
    trd_rpt_dt < as.Date("2012-02-06")
  )

# Trades w/o cancelations
## match the orig_msg_seq_nb of the C-message
## to the msg_seq_nb of the main message
trace_pre_T <- trace_all |>
  filter(
    trc_st == "T",
    trd_rpt_dt < as.Date("2012-02-06")
  ) |>
  anti_join(trace_pre_C,
    by = c("cusip_id",
      "msg_seq_nb" = "orig_msg_seq_nb",
      "entrd_vol_qt", "rptd_pr", "rpt_side_cd",
```

```
      "cntra_mp_id", "trd_exctn_dt", "trd_exctn_tm"
    )
  )

# Corrections (trc_st = W) - W can also correct a previous W
trace_pre_W <- trace_all |>
  filter(
    trc_st == "W",
    trd_rpt_dt < as.Date("2012-02-06")
  )

# Implement corrections in a loop
## Correction control
correction_control <- nrow(trace_pre_W)
correction_control_last <- nrow(trace_pre_W)

## Correction loop
while (correction_control > 0) {
  # Corrections that correct some msg
  trace_pre_W_correcting <- trace_pre_W |>
    semi_join(trace_pre_T,
      by = c("cusip_id", "trd_exctn_dt",
        "orig_msg_seq_nb" = "msg_seq_nb"
      )
    )

  # Corrections that do not correct some msg
  trace_pre_W <- trace_pre_W |>
    anti_join(trace_pre_T,
      by = c("cusip_id", "trd_exctn_dt",
        "orig_msg_seq_nb" = "msg_seq_nb"
      )
    )

  # Delete msgs that are corrected and add correction msgs
  trace_pre_T <- trace_pre_T |>
    anti_join(trace_pre_W_correcting,
      by = c("cusip_id", "trd_exctn_dt",
        "msg_seq_nb" = "orig_msg_seq_nb"
      )
    ) |>
    union_all(trace_pre_W_correcting)

  # Escape if no corrections remain or they cannot be matched
```

```
  correction_control <- nrow(trace_pre_W)
  if (correction_control == correction_control_last) {
    correction_control <- 0
  }
  correction_control_last <- nrow(trace_pre_W)
}


# Clean reversals
## Record reversals
trace_pre_R <- trace_pre_T |>
  filter(asof_cd == "R") |>
  group_by(
    cusip_id, trd_exctn_dt, entrd_vol_qt,
    rptd_pr, rpt_side_cd, cntra_mp_id
  ) |>
  arrange(trd_exctn_tm, trd_rpt_dt, trd_rpt_tm) |>
  mutate(seq = row_number()) |>
  ungroup()

## Remove reversals and the reversed trade
trace_pre <- trace_pre_T |>
  filter(is.na(asof_cd) | !(asof_cd %in% c("R", "X", "D"))) |>
  group_by(
    cusip_id, trd_exctn_dt, entrd_vol_qt,
    rptd_pr, rpt_side_cd, cntra_mp_id
  ) |>
  arrange(trd_exctn_tm, trd_rpt_dt, trd_rpt_tm) |>
  mutate(seq = row_number()) |>
  ungroup() |>
  anti_join(trace_pre_R,
    by = c(
      "cusip_id", "trd_exctn_dt", "entrd_vol_qt",
      "rptd_pr", "rpt_side_cd", "cntra_mp_id", "seq"
    )
  ) |>
  select(-seq)


# Agency trades -----------------------------------------------------------
# Combine pre and post trades
trace_clean <- trace_post |>
  union_all(trace_pre)
```

```r
# Keep angency sells and unmatched agency buys
## Agency sells
trace_agency_sells <- trace_clean |>
  filter(
    cntra_mp_id == "D",
    rpt_side_cd == "S"
  )

# Agency buys that are unmatched
trace_agency_buys_filtered <- trace_clean |>
  filter(
    cntra_mp_id == "D",
    rpt_side_cd == "B"
  ) |>
  anti_join(trace_agency_sells,
    by = c(
      "cusip_id", "trd_exctn_dt",
      "entrd_vol_qt", "rptd_pr"
    )
  )

# Agency clean
trace_clean <- trace_clean |>
  filter(cntra_mp_id == "C") |>
  union_all(trace_agency_sells) |>
  union_all(trace_agency_buys_filtered)


# Additional Filters ------------------------------------------------------
trace_add_filters <- trace_clean |>
  mutate(days_to_sttl_ct2 = stlmnt_dt - trd_exctn_dt) |>
  filter(
    is.na(days_to_sttl_ct) | as.numeric(days_to_sttl_ct) <= 7,
    is.na(days_to_sttl_ct2) | as.numeric(days_to_sttl_ct2) <= 7,
    wis_fl == "N",
    is.na(spcl_trd_fl) | spcl_trd_fl == "",
    is.na(asof_cd) | asof_cd == ""
  )


# Output ------------------------------------------------------------------
# Only keep necessary columns
trace_final <- trace_add_filters |>
  arrange(cusip_id, trd_exctn_dt, trd_exctn_tm) |>
```

```
    select(
      cusip_id, trd_exctn_dt, trd_exctn_tm,
      rptd_pr, entrd_vol_qt, yld_pt, rpt_side_cd, cntra_mp_id
    ) |>
    mutate(trd_exctn_tm = format(as_datetime(trd_exctn_tm), "%H:%M:%S"))

  # Return
  return(trace_final)
}
```

Bibliography

Abadie, A., Athey, S., Imbens, G. W., and Wooldridge, J. (2017). When should you adjust standard errors for clustering? *Working Paper*.

Allaire, J. and Chollet, F. (2022). *keras: R interface to 'Keras'*. R package version 2.9.0.

Areal, N. (2021). *frenchdata: Download data sets from Kenneth's French finance data library site*. R package version 0.2.0.

Armstrong, W., Eddelbuettel, D., and Laing, J. (2022). *Rblpapi: R Interface to 'Bloomberg'*. R package version 0.3.13.

Arnott, R., Harvey, C. R., and Markowitz, H. (2019). A backtesting protocol in the era of machine learning. *The Journal of Financial Data Science*, 1(1):64–74.

Avramov, D., Cheng, S., and Metzker, L. (2022a). Machine learning vs. economic restrictions: Evidence from stock return predictability. *Management Science (forthcoming)*.

Avramov, D., Cheng, S., Metzker, L., and Voigt, S. (2022b). Integrating factor models. *The Journal of Finance (forthcoming)*.

Bai, J., Bali, T. G., and Wen, Q. (2019). Common risk factors in the cross-section of corporate bond returns. *Journal of Financial Economics*, 131(3):619–642.

Balduzzi, P. and Lynch, A. W. (1999). Transaction costs and predictability: Some utility cost calculations. *Journal of Financial Economics*, 52(1):47–78.

Balduzzi, P. and Lynch, A. W. (2000). Predictability and transaction costs: The impact on rebalancing rules and behavior. *The Journal of Finance*, 55(5):2285–2309.

Bali, T. G., Engle, R. F., and Murray, S. (2016). *Empirical asset pricing: The cross section of stock returns*. John Wiley & Sons.

Ball, R. (1978). Anomalies in relationships between securities' yields and yield-surrogates. *Journal of Financial Economics*, 6(2–3):103–126.

Banz, R. W. (1981). The relationship between return and market value of common stocks. *Journal of Financial Economics*, 9(1):3–18.

Bergé, L. (2018). Efficient estimation of maximum likelihood models with multiple fixed-effects: The R package FENmlm. *CREA Discussion Papers*. R package version 0.10.4.

Bessembinder, H., Kahle, K. M., Maxwell, W. F., and Xu, D. (2008). Measuring abnormal bond performance. *Review of Financial Studies*, 22(10):4219–4258.

Bessembinder, H., Maxwell, W., and Venkataraman, K. (2006). Market transparency, liquidity externalities, and institutional trading costs in corporate bonds. *Journal of Financial Economics*, 82(2):251–288.

Black, F. and Scholes, M. (1973). The pricing of options and corporate liabilities. *Journal of Political Economy*, 81(3):637–654.

Boudt, K., Cornelissen, J., Payseur, S., Kleen, O., and Sjoerup, E. (2022). *highfrequency: Tools for highfrequency data analysis*. R package version 0.9.5.

Boysel, S. and Vaughan, D. (2021). *fredr: An R client for the 'FRED' API*. R package version 2.1.0.

Brandt, M. W. (2010). Portfolio choice problems. In Ait-Sahalia, Y. and Hansen, L. P., editors, *Handbook of Financial Econometrics: Tools and Techniques*, volume 1 of *Handbooks in Finance*, pages 269–336. North-Holland.

Brandt, M. W., Santa-Clara, P., and Valkanov, R. (2009). Parametric portfolio policies: Exploiting characteristics in the cross-section of equity returns. *Review of Financial Studies*, 22(9):3411–3447.

Brown, S. J. (1976). *Optimal portfolio choice under uncertainty: A Bayesian approach*. Phd thesis, University of Chicago.

Bryan, J. (2022). *Happy Git and GitHub for the useR*.

Bryzgalova, S., Pelger, M., and Zhu, J. (2022). Forest through the trees: Building cross-sections of stock returns. *Working Paper*.

Cameron, A. C., Gelbach, J. B., and Miller, D. L. (2011). Robust inference with multiway clustering. *Journal of Business & Economic Statistics*, 29(2):238–249.

Campbell, J. Y. (1987). Stock returns and the term structure. *Journal of Financial Economics*, 18(2):373–399.

Campbell, J. Y., Hilscher, J., and Szilagyi, J. (2008). In search of distress risk. *The Journal of Finance*, 63(6):2899–2939.

Campbell, J. Y., Lo, A. W., MacKinlay, A. C., and Whitelaw, R. F. (1998). The econometrics of financial markets. *Macroeconomic Dynamics*, 2(4):559–562.

Campbell, J. Y. and Shiller, R. J. (1988). Stock prices, earnings, and expected dividends. *The Journal of Finance*, 43(3):661–676.

Campbell, J. Y. and Vuolteenaho, T. (2004). Inflation illusion and stock prices. *American Economic Review*, 94(2):19–23.

Campbell, J. Y. and Yogo, M. (2006). Efficient tests of stock return predictability. *Journal of Financial Economics*, 81(1):27–60.

Cara, C. (2021). *DatastreamDSWS2R: Provides a link between the 'Refinitiv Datastream' system and R*. R package version 1.8.2.

Chen, A. Y. and Zimmermann, T. (2022). Open source cross-sectional asset pricing. *Critical Finance Review*, 11(2):207–264.

Chen, H.-Y., Lee, A. C., and Lee, C.-F. (2015). Alternative errors-in-variables models and their applications in finance research. *The Quarterly Review of Economics and Finance*, 58:213–227.

Chen, L., Pelger, M., and Zhu, J. (2019). Deep learning in asset pricing. *Management Science (forthcoming)*.

Chopra, V. K. and Ziemba, W. T. (1993). The effect of errors in means, variances, and covariances on optimal portfolio choice. *Journal of Portfolio Management*, 19(2):6–11.

Cochrane, J. H. (2005). Writing tips for PhD students. *Note*.

Cochrane, J. H. (2009). *Asset pricing (revised edition)*. Princeton University Press.

Cochrane, J. H. (2011). Presidential address: Discount rates. *The Journal of Finance*, 66(4):1047–1108.

Coqueret, G. and Guida, T. (2020). *Machine learning for factor investing: R version*. Chapman and Hall/CRC.

D'Agostino McGowan, L. and Bryan, J. (2021). *googledrive: An interface to google drive*. R package version 2.0.0.

Dancho, M. and Vaughan, D. (2022a). *tidyquant: Tidy quantitative financial analysis*. R package version 1.0.5.

Dancho, M. and Vaughan, D. (2022b). *timetk: A tool kit for working with time series in R*. R package version 2.8.1.

Davis, M. H. A. and Norman, A. R. (1990). Portfolio selection with transaction costs. *Mathematics of Operations Research*, 15(4):676–713.

De Prado, M. L. (2018). *Advances in financial machine learning*. John Wiley & Sons.

DeMiguel, V., Martín-Utrera, A., and Nogales, F. J. (2015). Parameter uncertainty in multiperiod portfolio optimization with transaction costs. *Journal of Financial and Quantitative Analysis*, 50(6):1443–1471.

DeMiguel, V., Nogales, F. J., and Uppal, R. (2014). Stock return serial dependence and out-of-sample portfolio performance. *Review of Financial Studies*, 27(4):1031–1073.

Dick-Nielsen, J. (2009). Liquidity biases in TRACE. *The Journal of Fixed Income*, 19(2):43–55.

Dick-Nielsen, J. (2014). How to clean enhanced TRACE data. *Working Paper*.

Dick-Nielsen, J., Feldhütter, P., and Lando, D. (2012). Corporate bond liquidity before and after the onset of the subprime crisis. *Journal of Financial Economics*, 103(3):471–492.

Dixon, M. F., Halperin, I., and Bilokon, P. (2020). *Machine learning in finance*. Springer.

Donald, S. and Lang, K. (2007). Inference with difference-in-differences and other panel data. *The Review of Economics and Statistics*, 89(2):221–233.

Easley, D., de Prado, M., O'Hara, M., and Zhang, Z. (2020). Microstructure in the machine age. *Review of Financial Studies*, 34(7):3316–3363.

Edwards, A. K., Harris, L. E., and Piwowar, M. S. (2007). Corporate bond market transaction costs and transparency. *The Journal of Finance*, 62(3):1421–1451.

Erickson, T. and Whited, T. M. (2012). Treating measurement error in Tobin's q. *Review of Financial Studies*, 25(4):1286–1329.

Fama, E. F. and French, K. R. (1989). Business conditions and expected returns on stocks and bonds. *Journal of Financial Economics*, 25(1):23–49.

Fama, E. F. and French, K. R. (1992). The cross-section of expected stock returns. *The Journal of Finance*, 47(2):427–465.

Fama, E. F. and French, K. R. (1993). Common risk factors in the returns on stocks and bonds. *Journal of Financial Economics*, 33(1):3–56.

Fama, E. F. and French, K. R. (1997). Industry costs of equity. *Journal of Financial Economics*, 43(2):153–193.

Fama, E. F. and MacBeth, J. D. (1973). Risk, return, and equilibrium: Empirical tests. *Journal of Political Economy*, 81(3):607–636.

Fan, J., Fan, Y., and Lv, J. (2008). High dimensional covariance matrix estimation using a factor model. *Journal of Econometrics*, 147(1):186–197.

Fazzari, S. M., Hubbard, R. G., Petersen, B. C., Blinder, A. S., and Poterba, J. M. (1988). Financing constraints and corporate investment. *Brookings Papers on Economic Activity*, 1988(1):141–206.

Frazzini, A. and Pedersen, L. H. (2014). Betting against beta. *Journal of Financial Economics*, 111(1):1–25.

Gagliardini, P., Ossola, E., and Scaillet, O. (2016). Time-varying risk premium in large cross-sectional equity data sets. *Econometrica*, 84(3):985–1046.

Gareth, J., Daniela, W., Trevor, H., and Robert, T. (2013). *An introduction to statistical learning: With applications in R*. Spinger.

Garlappi, L., Uppal, R., and Wang, T. (2007). Portfolio selection with parameter and model uncertainty: A multi-prior approach. *Review of Financial Studies*, 20(1):41–81.

Gârleanu, N. and Pedersen, L. H. (2013). Dynamic trading with predictable returns and transaction costs. *The Journal of Finance*, 68(6):2309–2340.

Gehrig, T., Sögner, L., and Westerkamp, A. (2020). Making portfolio policies work. *Working Paper*.

Goldstein, I., Koijen, R. S. J., and Mueller, H. M. (2021). COVID-19 and its impact on financial markets and the real economy. *Review of Financial Studies*, 34(11):5135–5148.

Gomolka, M. (2021). *simfinapi: Accessing 'SimFin' data*. R package version 0.2.0.

Grolemund, G. and Wickham, H. (2011). Dates and times made easy with lubridate. *Journal of Statistical Software*, 40(3):1–25. R package version 1.8.0.

Gu, S., Kelly, B., and Xiu, D. (2020). Empirical asset pricing via machine learning. *Review of Financial Studies*, 33(5):2223–2273.

Gulen, H. and Ion, M. (2015). Policy uncertainty and corporate investment. *Review of Financial Studies*, 29(3):523–564.

Guo, H. (2006). On the out-of-sample predictability of stock market returns. *The Journal of Business*, 79(2):645–670.

Halling, M., Yu, J., and Zechner, J. (2021). Primary corporate bond markets and social responsibility. *Working Paper*.

Handler, L., Jankowitsch, R., and Pasler, A. (2022). The effects of ESG performance and preferences on us corporate bond prices. *Working Paper*.

Handler, L., Jankowitsch, R., and Weiss, P. (2021). Covenant prices of US corporate bonds. *Working Paper*.

Harvey, C. R., Liu, Y., and Zhu, H. (2016). ... and the cross-section of expected returns. *Review of Financial Studies*, 29(1):5–68.

Hasler, M. (2021). Is the value premium smaller than we thought? *Working Paper*.

Hastie, T., Tibshirani, R., and Friedman, J. (2009). *The elements of statistical learning: Data mining, inference and prediction*. Springer, 2 edition.

Hautsch, N., Kyj, L. M., and Malec, P. (2015). Do high-frequency data improve high-dimensional portfolio allocations? *Journal of Applied Econometrics*, 30(2):263–290.

Hautsch, N. and Voigt, S. (2019). Large-scale portfolio allocation under transaction costs and model uncertainty. *Journal of Econometrics*, 212(1):221–240.

Henry, L. and Wickham, H. (2020). *purrr: Functional programming tools*. R package version 0.3.4.

Henry, L. and Wickham, H. (2022). *rlang: Functions for base types and core R and 'Tidyverse' features*. R package version 1.0.6.

Hoerl, A. E. and Kennard, R. W. (1970). Ridge regression: Applications to nonorthogonal problems. *Technometrics*, 12(1):69–82.

Hornik, K. (1991). Approximation capabilities of multilayer feedforward networks. *Neural Networks*, 4(2):251–257.

Hou, K., Xue, C., and Zhang, L. (2014). Digesting anomalies: An investment approach. *Review of Financial Studies*, 28(3):650–705.

Hou, K., Xue, C., and Zhang, L. (2020). Replicating anomalies. *Review of Financial Studies*, 33(5):2019–2133.

Huang, J.-Z. and Shi, Z. (2021). What do we know about corporate bond returns? *Annual Review of Financial Economics*, 13(1):363–399.

Hull, J. C. (2020). *Machine learning in business. An introduction to the world of data science*. Independently published.

Huynh, T. D. and Xia, Y. (2021). Climate change news risk and corporate bond returns. *Journal of Financial and Quantitative Analysis*, 56(6):1985–2009.

Ibrahim, M. (2021). *Riex: IEX stocks and market data*. R package version 1.0.2.

Jacobsen, B. (2014). Some research and writing tips. *Note*.

Jagannathan, R. and Ma, T. (2003). Risk reduction in large portfolios: Why imposing the wrong constraints helps. *The Journal of Finance*, 58(4):1651–1684.

Jegadeesh, N. and Titman, S. (1993). Returns to buying winners and selling losers: Implications for stock market efficiency. *The Journal of Finance*, 48(1):65–91.

Jensen, T. I., Kelly, B. T., Malamud, S., and Pedersen, L. H. (2022a). Machine learning and the implementable efficient frontier. *Working Paper*.

Jensen, T. I., Kelly, B. T., and Pedersen, L. H. (2022b). Is there a replication crisis in finance? *The Journal of Finance (forthcoming)*.

Jiang, J., Kelly, B. T., and Xiu, D. (2022). (Re-)Imag(in)ing Price Trends. *The Journal of Finance (forthcoming)*.

Jobson, D. J. and Korkie, B. (1980). Estimation for Markowitz efficient portfolios. *Journal of the American Statistical Association*, 75(371):544–554.

Jorion, P. (1986). Bayes-Stein estimation for portfolio analysis. *Journal of Financial and Quantitative Analysis*, 21(03):279–292.

Kan, R. and Zhou, G. (2007). Optimal portfolio choice with parameter uncertainty. *Journal of Financial and Quantitative Analysis*, 42(3):621–656.

Kearney, M. W. (2019). rtweet: Collecting and analyzing twitter data. *Journal of Open Source Software*, 4(42):1829. R package version 0.7.0.

Kelly, B. T., Palhares, D., and Pruitt, S. (2021). Modeling corporate bond returns. *Working Paper*.

Kiesling, L. (2003). Writing tips for economics (and pretty much anything else). *Note*.

Kim, D. (1995). The errors in the variables problem in the cross-section of expected stock returns. *The Journal of Finance*, 50(5):1605–1634.

Kleen, O. (2021). *alfred: Downloading time series from ALFRED database for various vintages*. R package version 0.2.0.

Kothari, S. P. and Shanken, J. A. (1997). Book-to-market, dividend yield, and expected market returns: A time-series analysis. *Journal of Financial Economics*, 44(2):169–203.

Kuhn, M. and Silge, J. (2018). *Tidy modeling with R*. John Wiley & Sons.

Kuhn, M. and Wickham, H. (2020). *Tidymodels: a collection of packages for modeling and machine learning using tidyverse principles*. R package version 1.0.0.

Lamont, O. (1998). Earnings and expected returns. *The Journal of Finance*, 53(5): 1563–1587.

Ledoit, O. and Wolf, M. (2003). Improved estimation of the covariance matrix of stock returns with an application to portfolio selection. *Journal of Empirical Finance*, 10(5):603–621.

Ledoit, O. and Wolf, M. (2004). Honey, I shrunk the sample covariance matrix. *The Journal of Portfolio Management*, 30(4):110–119.

Ledoit, O. and Wolf, M. (2012). Nonlinear shrinkage estimation of large-dimensional covariance matrices. *The Annals of Statistics*, 40(2):1024–1060.

Lintner, J. (1965). Security prices, risk, and maximal gains from diversification. *The Journal of Finance*, 20(4):587–615.

Magill, M. J. P. and Constantinides, G. M. (1976). Portfolio selection with transactions costs. *Journal of Economic Theory*, 13(2):245–263.

Markowitz, H. (1952). Portfolio selection. *The Journal of Finance*, 7(1):77–91.

Massicotte, P. and Eddelbuettel, D. (2022). *gtrendsR: Perform and display Google trends queries*. R package version 1.5.1.

Mclean, R. D. and Pontiff, J. (2016). Does academic research destroy stock return predictability? *The Journal of Finance*, 71(1):5–32.

McTaggart, R., Daroczi, G., and Leung, C. (2021). *Quandl: API wrapper for Quandl.com*. R package version 2.11.0.

Menkveld, A. J., Dreber, A., Holzmeister, F., Huber, J., Johannesson, M., Kirchler, M., Neusüss, S., Razen, M., and Weitzel, U. (2021). Non-standard errors. *Working Paper*.

Merton, R. C. (1972). An analytic derivation of the efficient portfolio frontier. *Journal of Financial and Quantitative Analysis*, 7(4):1851–1872.

Mossin, J. (1966). Equilibrium in a capital asset market. *Econometrica*, 34(4):768–783.

Mullainathan, S. and Spiess, J. (2017). Machine learning: An applied econometric approach. *Journal of Economic Perspectives*, 31(2):87–106.

Müller, K. and Wickham, H. (2022). *tibble: Simple data frames*. R package version 3.1.8.

Müller, K., Wickham, H., James, D. A., and Falcon, S. (2022). *RSQLite: SQLite interface for R*. R package version 2.2.17.

Nagel, S. (2021). *Machine learning in asset pricing*. Princeton University Press.

Newey, W. K. and West, K. D. (1987). A simple, positive semi-definite, heteroskedasticity and autocorrelation consistent covariance Matrix. *Econometrica*, 55(3):703–708.

Newey, W. K. and West, K. D. (1994). Automatic lag selection in covariance matrix estimation. *The Review of Economic Studies*, 61(4):631–653.

O'Hara, M. and Zhou, X. A. (2021). Anatomy of a liquidity crisis: Corporate bonds in the COVID-19 crisis. *Journal of Financial Economics*, 142(1):46–68.

Persson, E. (2021). *ecb: Programmatic Access to the European Central Bank's Statistical Data Warehouse*. R package version 0.4.0.

Peters, R. H. and Taylor, L. A. (2017). Intangible capital and the investment-q relation. *Journal of Financial Economics*, 123(2):251–272.

Petersen, M. A. (2008). Estimating standard errors in finance panel data sets: Comparing approaches. *Review of Financial Studies*, 22(1):435–480.

Pflug, G., Pichler, A., and Wozabal, D. (2012). The 1/N investment strategy is optimal under high model ambiguity. *Journal of Banking & Finance*, 36(2):410–417.

Puhr, H. and Müllner, J. (2021). Let me Google that for you: Capturing globalization using Google Trends. *Working Paper*. R package version 0.0.12.

R Core Team (2022). R: A language and environment for statistical computing. R version 4.2.1 (2022-06-23, Funny-Looking Kid).

Ram, K. and Wickham, H. (2018). *wesanderson: A Wes Anderson palette generator*. R package version 0.3.6.

Regenstein Jr, J. K. (2018). *Reproducible finance with R: Code flows and shiny apps for portfolio analysis*. Chapman and Hall/CRC.

Roberts, M. R. and Whited, T. M. (2013). Endogeneity in empirical corporate finance. In *Handbook of the Economics of Finance*, volume 2, pages 493–572. Elsevier.

Robinson, D., Hayes, A., and Couch, S. (2022). *broom: Convert statistical objects into tidy tibbles*. R package version 1.0.1.

Seltzer, L. H., Starks, L., and Zhu, Q. (2022). Climate regulatory risk and corporate bonds. *Working Paper*.

Shanken, J. (1992). On the estimation of beta-pricing models. *Review of Financial Studies*, 5(1):1–33.

Sharpe, W. F. (1964). Capital asset prices: A theory of market equilibrium under conditions of risk . *The Journal of Finance*, 19(3):425–442.

Silge, J., Chow, F., Kuhn, M., and Wickham, H. (2022). *rsample: General resampling infrastructure*. R package version 1.0.0.

Simon, N., Friedman, J., Hastie, T., and Tibshirani, R. (2011). Regularization paths for Cox's proportional hazards model via coordinate descent. *Journal of Statistical Software*, 39(5):1–13. R package version 4.1-4.

Soebhag, A., Van Vliet, B., and Verwijmeren, P. (2022). Mind your sorts. *Working Paper*.

Stoeckl, S. (2022). *crypto2: Download crypto currency data from 'CoinMarketCap' without 'API'*. R package version 1.4.4.

Tibshirani, R. (1996). Regression shrinkage and selection via the LASSO. *Journal of the Royal Statistical Society. Series B (Methodological)*, 58(1):267–288.

Tsay, R. S. (2010). *Analysis of financial time series*. John Wiley & Sons.

Turlach, B. A., Weingessel, A., and Moler, C. (2019). *quadprog: Functions to solve quadratic programming problems*. R package version 1.5-8.

Varadhan, R. (2022). *alabama: Constrained nonlinear optimization*. R package version 2022.4-1.

Vaughan, D. (2021). *slider: Sliding window functions*. R package version 0.2.2.

Vaughan, D. and Dancho, M. (2022). *furrr: Apply mapping functions in parallel using futures*. R package version 0.3.1.

Vaughan, D. and Kuhn, M. (2022). *hardhat: Construct modeling packages*. R package version 1.2.0.

Waldstein, M. J. (2021). *edgarWebR: SEC filings access*. R package version 1.1.0.

Walter, D., Weber, R., and Weiss, P. (2022). Non-standard errors in portfolio sorts. *Working Paper*.

Wang, Z. (2005). A shrinkage approach to model uncertainty and asset allocation. *Review of Financial Studies*, 18(2):673–705.

Wasserstein, R. L. and Lazar, N. A. (2016). The ASA Statement on p-Values: Context, process, and purpose. *The American Statistician*, 70(2):129–133.

Welch, I. and Goyal, A. (2008). A comprehensive look at the empirical performance of equity premium prediction. *Review of Financial Studies*, 21(4):1455–1508.

Wickham, H. (2014). Tidy data. *Journal of Statistical Software*, 59(1):1–23.

Wickham, H. (2016). *ggplot2: Elegant graphics for data analysis*. Springer-Verlag New York. R package version 3.3.6.

Wickham, H. (2019). *stringr: Simple, consistent wrappers for common string operations*. R package version 1.4.1.

Wickham, H. (2021). *forcats: Tools for working with categorical variables (Factors)*. R package version 0.5.2.

Wickham, H., Averick, M., Bryan, J., Chang, W., McGowan, L. D., François, R., Grolemund, G., Hayes, A., Henry, L., Hester, J., Kuhn, M., Pedersen, T. L., Miller, E., Bache, S. M., Müller, K., Ooms, J., Robinson, D., Seidel, D. P., Spinu, V., Takahashi, K., Vaughan, D., Wilke, C., Woo, K., and Yutani, H. (2019). Welcome to the Tidyverse. *Journal of Open Source Software*, 4(43):1686. R package version 1.3.2.

Wickham, H. and Bryan, J. (2022). *readxl: Read excel files*. R package version 1.4.1.

Wickham, H., François, R., Henry, L., and Müller, K. (2022a). *dplyr: A grammar of data manipulation*. R package version 1.0.10.

Wickham, H. and Girlich, M. (2022). *tidyr: Tidy messy data*. R package version 1.2.1.

Wickham, H., Girlich, M., and Ruiz, E. (2022b). *dbplyr: A 'dplyr' back end for databases*. R package version 2.2.1.

Wickham, H. and Grolemund, G. (2016). *R for data science: Import, tidy, transform, visualize, and model data*. O'Reilly.

Wickham, H., Hester, J., and Bryan, J. (2022c). *readr: Read rectangular text data*. R package version 2.1.2.

Wickham, H., Hester, J., Chang, W., and Bryan, J. (2022d). *devtools: Tools to make developing R packages easier*. R package version 2.4.4.

Wickham, H., Ooms, J., and Müller, K. (2022e). *RPostgres: Rcpp interface to PostgreSQL*. R package version 1.4.4.

Wickham, H. and Seidel, D. (2022). *scales: Scale functions for visualization*. R package version 1.2.1.

Wilkinson, L. (2012). *The grammar of graphics*. Springer.

Wooldridge, J. M. (2010). *Econometric analysis of cross section and panel data*. The MIT Press.

Wright, M. N. and Ziegler, A. (2017). ranger: A fast implementation of random forests for high dimensional data in C++ and R. *Journal of Statistical Software*, 77(1):1–17. R package version 0.14.1.

Xie, Y. (2016). *bookdown: Authoring books and technical documents with R Markdown*. Chapman and Hall/CRC.

Xie, Y., Allaire, J., and Grolemund, G. (2018). *R Markdown: The definitive guide*. Chapman and Hall/CRC.

Xie, Y., Dervieux, C., and Riederer, E. (2020). *R markdown cookbook*. Chapman and Hall/CRC.

Zaffaroni, P. and Zhou, G. (2022). Asset pricing: Cross-section predictability. *Working Paper*.

Zeileis, A. (2006). Object-Oriented computation of sandwich estimators. *Journal of Statistical Software*, 16(9):1–16. R package version 3.0-2.

Zeileis, A. and Hothorn, T. (2002). Diagnostic checking in regression relationships. *R News*, 2(3):7–10. R package version 0.9-40.

Zou, H. and Hastie, T. (2005). Regularization and variable selection via the elastic net. *Journal of the Royal Statistical Society. Series B (Statistical Methodology)*, 67(2):301–320.

Index

Alpha, 91
AMEX, 101
API, 23

Backtesting, 205, 217
Beta, 71, 88, 128, 129
Big data, 46, 67
Book equity, 48, 112, 120
Book-to-market ratio, 112, 120, 129
Bookdown, xv
Breakpoints, 89, 104, 121

CAPM, 71, 91, 129, 157
Cash flows, 136
Colophon, xv
Corporate bonds, 55
Covariance, 15, 207
Covid 19, 13
Cryptocurrency, 67
Curly-curly, 90, 114
CUSIP, 56, 227

Data
- Bloomberg, 66
- Compustat, 47, 112, 119, 128, 136
- CPI, 28, 41, 42
- CRSP, 36, 71, 88, 100, 111, 119, 128, 136, 193, 194
- Crsp-Compustat Merged, 49
- Dow Jones Index, 9
- ECB, 66
- Eikon, 66
- Fama-French factors, 24, 71, 88, 100, 111, 162, 193
- FISD, 56, 146
- FRED, 28, 65
- Google, 67
- Industry portfolios, 25, 162, 205
- Linking table, 49
- Macro predictors, 26, 162
- Nested, 76
- q-factors, 25, 162
- SP 500, 12
- TAQ, 67
- TRACE, 59, 146
- YahooFinance, 4

Database, 23
- Cleaning, 33
- Connection, 32
- Creation, 30
- Fetch, 31
- Management, 33
- PostgreSQL, 36
- Remote connection, 30
- Schema, 52
- SQLite, 29

Dick-Nielsen cleaning, 227
Difference in differences, 145
Diversification, 16

Efficient frontier, 14, 17
Efficient portfolio, 17, 205
Error message, 14
Exchange
- AMEX, 41, 50
- Exchange codes, 39
- NASDAQ, 9, 41, 50
- NYSE, 41, 50

Factor
- Market, 24
- q-factors, 25
- Size, 24
- Value, 24

Factor model, 157

Factor zoo, 157
Fama-MacBeth, 127
Firm size, 100, 112
For-loops, 46
Functional programming, 90

Generative art, 223
ggplot2 theme, xv
Github, xi
  Gist, 59, 227
Google Drive, 26
Google trends, 67
Graph
  Area graph, 101
  Bar chart, 92, 109, 214
  Box plot, 80, 163
  Comparative statics, 210
  Diff-in-diff graph, 150, 151
  Efficient frontier, 18
  Heat map, 176
  Histogram, 6
  ML prediction path, 170
  Prediction error, 188
  Time series, 4
Gvkey, 48, 49

High-frequency data, 67

Industry codes, 39

Kenneth French homepage, 24
keras, 185

Lag observations, 5
Liquidity, 62
Long-short, 89

Machine learning, 179
Market capitalization, 38, 100, 112, 129
Markowitz optimization, 14
Minimum variance portfolio, 16, 206
Missing value, 5
Model uncertainty, 207
Momentum, 194
MSE, 159

NASDAQ, 101
Neural network, 179
NYSE, 101

Optimization, 205, 213
  Constrained, 211
Option pricing, 179
Out-of-sample, 217

P-hacking, 107
Parallel trends assumption, 150
Parallelization, 77
Paris (Climate) Agreement, 60, 145
Performance evaluation, 91, 198, 217
Permno, 37, 49
Pipe, xiii
Portfolio choice, 14, 205
Portfolio sorts, 121
  Dependent bivariate, 116
  Independent bivariate, 115
  Univariate, 87, 104
Power utility, 198
Preferred stock, 48

Random forests, 179
Rating, 64
Regression, 166
  Cross-section, 129
  Fama-MacBeth, 127
  Fixed effects, 138, 147
  Investment, 136
  Lasso, 160
  OLS, 158
  Panel, 136, 149
  Ridge, 159
  Yields, 149
Regulation T, 213
Return volatility, 14
Returns, 4, 6, 38
  Bonds, 147
  Delisting, 39
  Excess, 40
Risk aversion, 95
Risk-free rate, 24, 40
Robustness tests, 107, 143, 154
Rolling-window estimation, 73